Webb Society Deep–Sky Observer's Handbook
Volume 2: Planetary and Gaseous Nebulae

Webb Society Deep–Sky Observer's Handbook

Volume 2
Planetary and Gaseous Nebulae

Compiled by the Webb Society
Editor: Kenneth Glyn Jones, F.R.A.S.
Collaborating Authors:
David A. Allen, Ph.D., F.R.A.S.
Edmund S. Barker, F.R.A.S.

Enslow Publishers
Hillside, New Jersey 07205

1979

To Professor B. A. Vorontsov–Velyaminov
Pioneer in the study of Planetary Nebulae

First joint American-British edition, 1979

First published in the U.K. in 1978 as
The Webb Society Observers Handbook.
Volume II, Planetary Nebulae Gaseous Nebulae

Library of Congress Cataloging in Publication Data

Webb Society.
Webb Society deep–sky observer's handbook.

First published in 1975– under title: The Webb Society observers handbook.
Includes bibliographies.
CONTENTS: v. 1. Double stars.—v. 2. Planetary and gaseous nebulae.
1. Astronomy—Observers' manuals. I. Jones, Kenneth Glyn. II. Title.

QB64.W35 1979 523 78-31260
In the U.S.A.: ISBN 0-89490-028-5 (v. 2)

Manufactured in the United States of America
10 9 8 7

CONTENTS

LIST OF ILLUSTRATIONS.

General Preface

Named after the Rev. T.W. Webb (1807-1885), an eminent amateur astronomer and author of the classic Celestial Objects for Common Telescopes, the Webb Society exists to encourage the study of double stars and deep-sky objects. It has members in almost every country where amateur astronomy flourishes. It has a number of sections, each under a director with wide experience in the particular field, the main ones being double stars, nebulae and clusters, minor planets, supernova watch, and astro-photography. Publications include a Quarterly Journal containing articles and special features, book reviews, and section reports that cover the society's activities. Membership is open to anyone whose interests are compatible. Application forms and answers to queries are available from Dr. G.S. Whiston, Secretary, Webb Society, 'Chestnuts', 1 Cramhurst Drive, Witley, Surrey, England.

Webb's Celestial Objects for Common Telescopes, first published in 1859, must have been among the most popular books of its kind ever written. Running through six editions by 1917, it still is in print although the text is of more historical than practical interest to the amateur of today. Not only has knowledge of the universe been transformed totally by modern developments, but the present generation of amateur astronomers has telescopes and other equipment that even the professional of Webb's day would have envied.

The aim of the new Webb Society Deep-Sky Observer's Handbook is to provide a series of observer's manuals that do justice to the equipment that is available today and to cover fields that have not been adequately covered by other organizations of amateurs. We have endeavored to make these guides the best of their kind: they are written by experts, some of them professional astronomers, who have had considerable practical experience with the problems and pleasures of the amateur astronomer. The manuals can be used profitably by the beginner, who will find much to stimulate his enthusiasm and imagination. However, they are designed primarily for the more experienced amateur who seeks greater scope for the exercise of his skills.

Each handbook volume is complete with respect to its subject. The reader is given an adequate historical and theoretical basis for a modern understanding of the physical role of the objects covered in the wider context of the universe. He is provided with a thorough exposition of observing methods, including the construction and operation of ancillary equipment such as micrometers and simple spectroscopes. Each volume contains a detailed and comprehensive catalogue of objects for the amateur to locate--and to observe with an eye made more perceptive by the knowledge he has gained.

We hope that these volumes will enable the reader to extend his abilities, to exploit his telescope to its limit, and to tackle the challenging difficulties of new fields of observation with confidence of success.

Preface
Volume 2: Planetary and Gaseous Nebulae

With this volume we move into the region of what amateurs loosely, but aptly term, 'deep-sky' observing. The two classes of object dealt with here; Planetary Nebulae and Gaseous Nebulae, each offer in their different ways, a considerable challenge to the enthusiastic observer.

Planetary nebulae, apart from one or two notable exceptions such as the 'Dumb-bell' Nebula (M27) and the 'Ring' Nebula (M57), are either bright but extremely small or large but extremely faint: some, in fact are both very faint and very small, but yet can be detected in telescopes of quite moderate aperture by applying carefully the methods described in this volume.

Gaseous nebulae can vary from the immediately spectacular - as in the case of the 'Great Orion Nebula' (M42) - to the beautiful but elusive 'Veil' Nebula in Cygnus, but even these familiar objects can yield new delights when examined with a more enlightened eye. And there are the too-often neglected Dark Nebulae, such as the 'Fish's Mouth' in M42 or the famous 'Coal-Sack' in Crux, which can have a fascination all their own.

In this volume the reader is presented with an adequate descriptive text, comprising a brief historical background, a sound theoretical resume and detailed notes on practical observing methods and practice. In addition, a selected catalogue is provided for each of the two classes of object under review. The observing notes and numerous drawings have been compiled exclusively from those contributed by members of the Webb Society's Nebulae and Clusters Section and detailed acknowledgements are given in the text.

The editor's task, I am glad to say, has been a relatively easy, and entirely pleasant one, for both text and catalogues have been written jointly by two of our leading Webb Society members, Dr. D.A. Allen and Edmund Barker: their work has been virtually unmarred by editorial interference.

Dave Allen, who is a founder-member of the Society, is a professional astronomer specialising in infra-red astronomy, who also maintains an enthusiastic 'amateur' interest in visual astronomical observation. He graduated at Cambridge in 1967 and subsequently spent 17 months as a Carnegie Fellow at Hale Observatories followed by 2½ years at Royal Greenwich Observatory. Since obtaining his Ph.D. in infra-red astronomy in 1971 he has travelled widely and has observed with many of the world's largest telescopes. He is at present enjoying a S.R.C. Fellowship at the Anglo Australian Telescope. He is the author of 'Infra-red: The New Astronomy' (David and Charles, distributors, 1975) and has had some 50 papers published in various scientific journals.

Ed Barker, by profession, is an artist who paints compelling abstract compositions for satisfaction and evocative landscapes for bread-and-butter, but astronomy competes strongly with art for his creative energies. He was the first recipient of the Webb Society Award (in 1973) and has been the energetic Director of the Nebulae and Clusters Section since 1974. In the pages of our Quarterly Journal he has demonstrated the effective use of an eye-piece prism as a visual spectroscope - a technique which has greatly extended the scope of the amateur, especially in identifying the smaller planetary nebulae. His draughtsmanship has been invaluable in the production of this Handbook, and the execution of its many diagrams and drawings are due to him.

As in the case of Volume I of this series (Double Stars) a great deal of the organization involved in the production of the Handbook has been in the capable hands of Eddie Moore, our Publications Officer. Eddie, now retired from the B.B.C., together with his observatory, has moved to Kent, where he finds observing conditions several magnitudes superior to those he endured in North West London.

Finally the Webb Society is greatly honoured by the acceptance of the dedication of this volume by Professor B.A. Vorontsov-Velyaminov of the Sternberg Astronomical Institute of Moscow University, who is not only eminent in astronomy generally, but whose authority in the study of planetary nebulae is unquestioned.

AUTHORS' ACKNOWLEDGEMENTS.

Some of the research which was undertaken for this Handbook was made possible by our access to the libraries of the Anglo-Australian Observatory, the Royal Greenwich Observatory and the Royal Astronomical Society.

GENERAL INTRODUCTION.

The present day astronomer, whether professional or amateur, has his antecedents in the observers of the late eighteenth and early nineteenth centuries. The observations undertaken by the Herschels, Lord Rosse, Lassell, Huggins and many others, form the basis of contemporary astronomy. As time progressed, the development of instrumental techniques led to the gradual widening of the gap between the professional and the amateur. The difference between the two, however, is only that of approach, the basic motive for both groups being true scientific enquiry.

Today what can the amateur with a predilection for Nebulae and Clusters observation achieve? The answer is that this depends very much upon the individual. Many amateurs will concentrate on visual work, others upon photographic observation. In both these approaches certain people will cover the whole range of objects accessible to their equipment, while others will feel the need to specialize, if not for all of the time, at least for a great part of it.

Beyond the realms of visual and photographic work there lie the fields of extra-specialization. J.B. Sidgwick, in 'The Amateur Astronomers' Handbook', commented upon the tendency of the amateur in Britain to be conservative in his methods, and compared him with the American amateur, who tends to show a more progressive attitude towards observational methods and equipment. Doubtless the situation is not much different now from the time when Sidgwick was writing, but there is no need for it to be. Not every amateur has the desire to explore the more intricate paths of extra-specialization, but for those who do, there is certainly nothing to lose by going ahead.

Certain work carried out professionally is within the capabilities of determined amateurs. Photoelectric photometry, photographic spectroscopy, Fabry-Perot Interferometry can all be practised by those who feel drawn to such activity. The present volumes, however, are concerned with visual observation only. The practical limits of the observer/telescope combination can always be extended, and with the neccessary self-criticism, satisfaction in one's work will become all the greater.

Unlike the observer of Double Stars, the Nebulae and Clusters observer has little auxiliary equipment with which to extend the accuracy of his observations. On the understanding that the limits of visual work are fully realised, however, much of interest can be accomplished.

While the general run of telescopes used by amateurs are in the 6 to 12-inch group, many people will own, or have access to, instruments of larger aperture. In the Webb Society observations with 14, 16½ and 18-inch telescopes are made, and there is the

General Introduction.

likelihood that more telescopes of similar size will be employed as time progresses.

The range of clusters, nebulae and galaxies available to the amateur is by no means limited to the entries in the NGC and IC. Many faint sources are within reach of quite moderate apertures, particularly objects of fairly small angular diameter. Although these may only show a stellar or quasi-stellar image, or (in the case of more extended sources) a barely discernable glow, many are worth listing in catalogues for the visual observer.

Details of the founding of the Webb Society are to be found in the introduction to the Double Star Handbook. It is to be hoped that the present volume, as well as the subsequent ones, will act as an incentive to many observers to extend the boundaries of their work, as well as giving some idea of the number and diversity of objects within the range of differing apertures.

Acknowledgement is due to the assistance of the following: Robert Argyle for computing precession figures and tracking down requested material; Jeffrey Perkins for supplying 1975 positions from the RNGC, and to Mrs. E. Lake of the R.A.S. Library for her help in dealing with many requests.

Finally, many thanks to George Whiston, Malcolm Thomson, Patrick Brennan, Steven Hynes and Scott Selleck for offers to assist with, and continued interest in, the progress of the Handbook. Thanks are also due to those members of the Society whose observations make up the catalogue sections of the respective volumes, and without whose efforts these volumes would not have appeared.

This second volume of the Webb Society Observers' Handbook will be concerned with visual observations of planetary nebulae and gaseous nebulae. Succeeding volumes will cover observations of open and globular clusters and galaxies. In all these cases some basic treatment of the properties, structure and evolutionary development of these objects is required, although we can do no more than summarize the more important aspects here.

PART ONE : PLANETARY NEBULAE.

INTRODUCTION.

The similarity of planetary nebulae to planets is entirely superficial, but does indicate one important feature of these emission nebulae: their regular outline. In contrast to the irregular and often streaky H II regions described in Part Two, planetary nebulae are tidy concise objects. Their profile is usually circular or slightly elliptical, because of their almost spherical proportions; in many cases the centre is darker than the rim, indicating a deficiency of material there. All have central stars that excite them, but in many cases these stars are extremely faint in the visible part of the spectrum. With surface temperatures up to 100,000°K, (cf. the sun at 5,000°K), virtually all their radiation occurs in the ultraviolet, and much of this is used in illuminating the nebula.

Why should such hot stars be found at the centres of spherical nebulae? This is a question which had long puzzled theoreticians, and it is only in the last ten years that an answer has been reached which looks moderately acceptible. Planetary nebulae, the model claims, evolve from cool red giant or supergiant stars.

In the late stages of stellar evolution, hydrogen becomes depleted in the core of a star because it has all been transformed into helium. For various reasons the temperature then rises until the helium begins to combine to form heavier elements. This process releases very large amounts of energy which inflate the star to gargantuan proportions. In its obesity the star can no longer hold itself together, so it begins a celestial slimming campaign, smoking off its outer layers. This process continues until all the hydrogen has been ejected, leaving the core behind. The hydrogen, expanding slowly as a spherical shell, forms the nebula, and the extremely hot core is its central star.

Because its central star is so hot, the emission spectrum is of characteristically high excitation - i.e., has lines of highly ionized atoms. For example, helium loses both its electrons and produces a line at 4686 Å.

These stars, the nuclei of planetary nebulae, are classed according to their spectral characteristics. Full details of the various classes can be found in the introduction to the catalogue of planetary nebulae, but on the subject of star types it is apposite to mention the similarity of certain nuclei to Wolf-Rayet stars. Although the absolute magnitudes of Wolf-Rayet type nuclei in planetary nebulae are lower than those of Wolf-Rayet stars proper, the spectral features are similar. For those Wolf-Rayet nuclei with carbon predominating in their spectra, the classification is WC, and for those with nitrogen in abundance it is WN. In those cases where these elements are present in equal amounts the star is classed as WC+WN. To these we can add nuclei of the type present in the nebula NGC 246, i.e., stars of very high excitation spectra with strong [OVI] emission lines. Objects of this type have been studied by Greenstein and Minkowski (Ap. J. 140, 1601, 1964).

While optically the nuclei of planetary nebulae appear to be single objects, evidence is accumulating that in many cases binary systems are present. Binary stars are possible in NGC 1514, H1-5, M1-2, K1-2 and IC 4406, and, in the case of NGC 1360 the central star is the spectroscopic binary CPD -26° 389.

A further aspect of planetary nebulae nuclei is variability, and it is of interest to note that certain nebulae of small angular diameters have been classed as variable stars, as in the case of Hb 6 (AS Sgr). The nucleus of the nebula He1-5 is the variable star FG Sge, and is certainly one of the most interesting of central stars. Since about the turn of the century this star has shown a steady increase in magnitude. In 1900 the magnitude was about 13.5, and by the mid 1960's this had risen to 9.5. Since 1955 data obtained has shown that the temperature of the star has decreased while spectral changes have, of course, occured also. In 1955 the star was classed as B4 I with P Cyg features while by 1967 the spectrum was A5 Ia, with no trace of the P Cyg features remaining.

Other nebulae in which variable nuclei are possible are the the following. NGCs 2346, 6891 and 7662, ICs 2149 and 4593 plus small objects such as M1-67, M3-18 and H3-29.

While dwelling on the subject of variable nuclei, mention must be made of the discovery of an emission variable star in Sge. Given the designation HM Sge, this object is considered to be an embryonic planetary nebula; certainly many of the spectral features correspond with those of planetary nebulae. If this is the case, then a unique opportunity is presented to study the formative period of a planetary nebula. Details of this object will be found in Appendix 4.

We have seen that the central stars of some nebulae are very faint. In some cases no nuclei are to be seen at all, but not every case of this kind denotes a very faint star. In some dense nebulae the material blots out the nucleus, and in such cases it is possible to deduce the physical characteristics of the star from the nebular spectra. The same deduction can be made in those cases where the star is unseen but is not surrounded by dense matter.

Introduction.

In these latter cases the unseen nuclei must be of very low luminosity; this is inferred from the fact that the inner parts of the nebulae are composed of quite faint material. NGCs 6565, 6741 and IC 2165 all show an oval ring, uniformly bright at the edges and faint at the centre. This faint inner material should allow a nucleus of quite low luminosity to be seen, and the fact that this is not the case indicates the extreme faintness of such nuclei.

In Table 1 a selection of nebulae with no visible nuclei is given; all are objects of small diameter apart from NGC 2440.

Table 1. Nebulae with no visible nuclei.

IC 2165	IC 4634	IC 4732	NGC 6833
NGC 2440	NGC 6537	IC 4776	IC 5117
J 900	NGC 6644	NGC 6741	IC 5217

Shell structure and related phenomena.

The study of the shells or envelopes of planetary nebulae can furnish data on a) the shell structure and b) internal and expansion velocities of the nebulosity.

Structurally, the shells of nebulae show varieties of form. From the simple morphology of stellar and planet-like (disk) nebulae, the variations become more complex; a ring that is superimposed on a disk (NGC 7662), a broken, irregular shell in ring form (NGC 246) and small, elliptical objects which are surrounded by thin rings in which are set one or more point-like condensations (J 320, NGC 6210).

The structure of the nebular shells can be equated with different wavelengths; this is known as stratification of radiation, meaning the concentration of the more highly ionized atoms towards the centre of the nebula, i.e., nearer to the exciting star. Thus in NGCs 1535 and 6309 [NeV] and HeII is concentrated towards the centre while HeI and [OII] occur in the outer regions.

How the stratification of radiation relates to the shell structure of nebulae can be seen from the following examples. IC 2165 is bipolar in the H lines and in [NII] (6548: 6584 Å) and this bipolarity is weak in [NeIV] (3869 Å) and disappears in [NeV] (3426 Å). In the latter two wavelengths the nebula shows as a ring and a disk respectively. Another object that shows identical aspects is NGC 7662.

From the above it can be appreciated that the angular diameters of the shells will vary according to the different wavelengths. Table 2 overleaf shows the dimensions in arcsec of three nebulae in four selected wavelengths.

Introduction.

Table 2. Monochromatic Images of Selected Nebulae.

Neb.	H beta	NeIII	HeII	NeV
NGC 2392	13.0	13.6	12.8	11.8
NGC 6816	17.6	19.2	14.4	10.8
NGC 7662	13.9	14.7	12.1	9.1

On plates taken with large telescopes, the shell structure of the larger planetary nebulae show great complexity. In the light of [OIII] M27 shows a fairly amorphous form, but in the H-alpha and [NII] wavelengths a great deal in the way of filamentary structure is evident; on a lesser scale the same is true of M97.

Many nebulae are at too great distances for their structural properties to be fully resolved, but in the case of a very large object such as NGC 7293 fine resolution can be obtained. In this nebula the ionization structure of H-alpha and [OIII] is very highly resolved, the H-alpha image recording as an annulus which is less clear in the [OIII] image. Minkowski (IAU Symp. 34, 1967) commented that if this nebula were bright enough to allow its photography in HeII it would probably show as a small disk.

The filamentary detail in the shell of NGC 7293 has been studied by Vorontsov-Velyaminov. These filaments are found to be about 1 to 0.6 arcsec in general width, giving actual sizes of the order of 150 by 3000 AU and masses of about 5×10^{25} g.

We come now to the motions of the shells of planetary nebulae; these are a) internal motions and b) expansion velocities. Unlike the violent ejection of gas in novae and supernovae, the nebular shells of planetary nebulae are in the process of slow expansion. Within the mean figures denoting expansion, different nebulae display variations in velocities, thus in the case of NGC 2392 the angular outward velocity decreases with increasing distance from the centre. In NGC 7009 the outer material, in the form of blobs, is moving outwards faster than the main material, but still shares in a well-defined linear relationship between angular velocity and nuclear distance. From this it would seem to indicate that all visible mass left the star at approximately the same time. However, the presence in this object of three distinct shells, (a feature it shares with some other nebulae), may indicate three relatively isolated ejection processes. In this context it must be mentioned that other nebulae show a double-shell structure, possibly indicating similar activity.

To show the variation in mean expansion velocities, we have selected six nebulae of different types and a fairly wide range of velocities, which are shown in Table 3 overleaf.

Table. 3. Expansion Velocities of Selected Nebulae.

Neb.	Vel.(km/sec)	Neb.	Vel.(km/sec)
IC 351	14.5	NGC 6543	12.0
NGC 1535	19.8	NGC 5657	17.9
IC 418	21.1	NGC 6818	28.3

When we come to the internal motions of nebulae we find that extremely complex activity is often the order. Aller has made studies of these motions in M27, and it is, of course, these larger and brighter objects that respond well to such investigations.

Finally, what is the overall picture of the evolution of the shell of a planetary nebula? For a start we can say that once the shell has actually begun to leave the regions of the nucleus, it will be a considerable time before this becomes apparent, the more so for very distant nebulae. We need only think of the large numbers of nebulae that are virtually indistinguishable from stars to see this. However, once the slow-moving shell attains respectable dimensions, then it will be apparent on photographs and eventually even visually.

Once the nebulosity takes on discernable structure, having departed from the pure disk phase, then differences become apparent. Not all nebulae show ring structure, but for ring nebulae proper, the following sequence is possible. As the apparent ring form expands, the uniformity of the ring, (seen in objects such as IC 418), decreases, and it shows as being less condensed at the ends of the major axis, (M57, NGC 6772). Eventually the two brighter arcs of material comprising the minor axis become virtually separated, only faint traces of linking matter remaining, structure that is seen in objects such as NGC 2474/5 and some Abell nebulae, assuming that not all of the latter are supernova remnants.

Distances.

In Appendix 3 we have considered the problems of distance determinations. There are, however, a few points which can be added on the subject relative to planetary nebulae. Although in the cases of relatively nearby nebulae proper motions can be used to help gain some indication of distances, these can only be applied to the nucleus of a nebula. Such cases are, in fact, few and far between, and probably only NGC 7293 is of use in this respect.

As no two nebulae have identical parameters, it is very difficult to arrive at a zero-point nebula which can be used as a standard candle. Although proper motions, nebular mass, angular expansion and interstellar extinction have been utilised to try and obtain meaningful distances, the fact that binary stars may form the nuclei of some nebulae may aid in more accurate distance determinations. For further reading on this most interesting and fundamental problem, reference is made to the following papers.

Shklovsky, I. Astr. J. Sov. Union. 33, 315, 1956.
O'Dell, C. Mon. Not. Roy. Astr. Soc. 132, 347, 1966.
Seaton, M. Ap. J. 135, 371, 1962.
Abell, G. Ap. J. 144, 259, 1966.

Classification.

The classification system for planetary nebulae first devised by Vorontsov-Velyaminov is still in use, although other attempts have been made to provide alternatives. Details can be found in the following.

Evans, D.S., Thackeray, A.D. Mon. Not. Roy. Astr. Soc. 110, 429, 1950.
Hromov, G.S. Sov. Astr. 6, 370, 1962.
Greig, W.E. Astr. Astrophys. 10, 161, 1971.

Full details of the Vorontsov-Velyaminov classification will be found in the introduction to the accompanying catalogue of planetary nebulae in Part 3 of this volume.

GENERAL HISTORY OF PLANETARY NEBULAE OBSERVATION.

The name 'planetary nebulae' is generally supposed to have been applied to these objects by William Herschel in 1785. Before this, however, in 1781, Darquier had already described M57 (a discovery of his) as 'looking like a fading planet'. Observations and discoveries of these objects continued until, in the 1860s, came a breakthrough in their observation due to spectroscopic studies of NGC 6543 by Huggins. This observation showed the object to have an emission spectrum, and was the first step in the realization that not all the nebulous objects observed over the years were comprised only of stars.

In the period following Huggins' observation, many more planetary nebulae were discovered, and as spectrographs became increasingly used, many objects of very small angular diameter were added to the number of known nebulae. Today the majority of planetary nebulae can be distinguished from stars only with very large telescopes and good seeing. To the amateur they appear stellar, and can be identified only by using a direct vision spectroscope, prism or finding chart. The discovery of planetary nebulae by spectroscopy began just before Dreyer compiled the NGC. Thus virtually all of the NGC planetaries are extended objects, while many of the IC and subsequently discovered nebulae are almost stellar.

Photographs of some planetary nebulae had been taken in the latter part of the nineteenth century by Isaac Roberts, but it was not until the latter part of World War I that Curtis, using the 36-inch Crossley reflector at Lick Observatory, made further detailed studies of these objects. Curtis also made numerous drawings from his plates, defining more clearly the distribution of material within certain nebulae.

In the early days of nebular spectroscopy, the brightest emission lines shown by planetary nebulae had been attributed to an element called nebulium. This purely hypothetical element was considered to occur only in gaseous nebulae. In the 1920s, I.S. Bowen, working at Mount Wilson and Palomar Observatories, showed that the strongest emission lines are the result of Oxygen atoms having lost two of their electrons. These, plus other lines of Oxygen, Nitrogen and Neon, are known as forbidden lines, due to the impossibility of their being duplicated in laboratory experiments on earth.

During the 1920s, Duncan, using the 60-inch and 100-inch reflectors on Mount Wilson, made further photographic studies of many gaseous nebulae, some of these being planetaries which Curtis had photographed earlier. With exposures of around two hours, Duncan found that many of the planetaries which Curtis had found to consist of only one envelope did, in fact, show two envelopes. Among the well-known objects of this type were M57, NGC 6826 and NGC 7662.

After World War II, further discoveries were made, notably by Minkowski (objective prism work) and by scrutiny of the Palomar Sky Survey plates. On the survey plates further nebulae, hitherto regarded as single-envelope objects, were found to exhibit an outer envelope, among them NGC 1514, NGC 4361, NGC 6369 and NGC 6554.

Among the new nebulae found on the survey plates were many of very low surface brightness and often large angular size. Although these nebulae, discovered by Abell, may not all be planetaries, some being probable supernova remnants, they are usually classed as planetaries in existing catalogues.

Planetary nebulae have also been found in some nearby galaxies, e.g., M31, NGC 147 and NGC 185, while new nebulae are currently being found on plates taken by the new telescopes in the southern hemisphere.

1. CATALOGUES OF PLANETARY NEBULAE.

The most comprehensive catalogue of planetary nebulae is that compiled by Perek and Kohoutek. It lists more than 1000 nebulae of which some 400 are stellar or very nearly so. In addition there are over 100 known in the Magellanic Clouds (several of them NGC objects) and a few are known in other nearby galaxies.

Unfortunately the Catalogue of Galactic Planetary Nebulae is an unreliable catalogue. Despite rejecting many objects once tentatively identified as planetary nebulae, it contains numerous objects which are certainly not. Some of the brightest of them are listed below in Table 4.

Table 4. Misclassified Planetary Nebulae.

NGC 1952	M1 Crab Nebula.
NGC 5189	H II region.
NGC 6302	H II region.
NGC 7027	Possibly compact H II region.
B13-11	G or Be star.
He2-10	Emission-line dwarf galaxy.
M1-11	Be star.
M1-15	Be star, probably in cluster.
H1-36	Symbiotic star.
K3-47	Me star.
K3-50	Compact H II region.
M4-18	Wolf-Rayet star.
H2-2	V455 Sco.
PC 18	AE Ara, Be star.
Sh2-266	H II region.

A list of the types of objects misclassified as planetaries, and the very simple reasons, is given below.

H II regions and nebulous clusters - because of emission and regular appearance.
Compact H II regions - because of spectrum.
Be stars and T Tauri stars - because of emission spectrum.
M stars and Mira variables - confusion in objective prism surveys.
G stars - at least one due to a faulty finding chart.
Novae - because of their spectra after maximum light.
Galaxies - because of their shape or their spectra.
Supernova remnants - because of their shape.

Since 1969 several lists of planetaries have been published. Covering the northern hemisphere, Kohoutek has given two further lists (1969, Bulletin Astr. Inst. Czechoslovakia, 20, 307, and 1972, Astron. Astrophys. 16, 291). The objects included are all extremely faint, and many are not planetaries at all.

Coverage of the southern hemisphere in the CGPN was poor, so many new discoveries can be expected. Some have already been made by Wray, (unpublished theses, but contained in Sanduleak, 1975, Publ. Warner and Swasey Obs., Vol. 2), while many more are showing up on plates taken in the Southern Schmidt Telescope surveys by the U.K. Schmidt Telescope Unit in Australia and the European Southern Observatory in Chile. Photographs of some of these nebulae can be found in: Longmore, Mon. Not. Roy. Astr. Soc., 178, 251-257, 1977.

A further publication of interest is a Bibliographical Index of Planetary Nebulae 1965-1976. This includes 148 new nebulae, bringing the known total to 1184, inclusive of misclassifications and uncertain cases. This index can be found in: Acker and Marcout, Astron. Astrophys. Suppl. 30, 217-221, 1977.

In the Atlas Coeli Catalogue, Becvar's list of planetaries is useful. It certainly includes all the bright NGC nebulae which are potentially within reach of modest amateur instruments, although some are so small that they would not normally be spotted without the aid of a good finding chart, direct vision spectroscope or prism. The list also includes a number of fainter objects, some of which are in the IC. For many no name is given, although a common name exists in all cases. There seems no particular pattern to Becvar's selection of objects; some extremely difficult nebulae being listed when other, easier objects are not. However, it is a difficult task to arrange planetaries into an 'ease of observing' order, and it would be churlish to criticize Becvar for his shortcomings in this regard.

A more serious criticism, however, is that of the listed magnitudes, which can be greatly misleading. The vexed question of the magnitudes of planetaries is aired overleaf, and it will suffice to note now that the magnitudes given in Becvar, although accurately measured, are of a heterogeneous nature; some differ from a meaningful brightness by one or two magnitudes.

2. VISUAL MAGNITUDES OF PLANETARY NEBULAE.

We are used to looking at stars with essentially continuous spectra, so that they emit almost equally strongly in two neighbouring wavelengths. Planetary nebulae (and H II regions) do not do this; all their light is emitted in narrow emission lines. The principal lines are:

Wavelength (Å)	Ion	Notes
6584	N II	Forbidden line.
6563	H I	H alpha.
6548	N II	Forbidden; = one third of 6584.
5007	O III	Forbidden; usually the strongest.
4959	O III	Forbidden; = one third of 5007.
4861	H I	H beta; = one third of H alpha.
4340	H I	H gamma; = 0.4 of H beta.
3727	O II	Forbidden; often strong.

Unlike stars, therefore, the apparent brightness of a planetary nebula depends upon what wavelengths, and hence which of these lines, the system responds to. Old photographic plates petered out to the red at about 4900 Å and so missed the strongest lines. Modern photometry on the UBV system also misses the strongest lines, B collecting only H beta and gamma, and V detecting a few weak lines plus 5007 very weakly at the edge of its bandpass. The old photographic or modern B or V magnitudes of planetary nebulae are generally too faint, yet it is these that are most often quoted.

What concerns us is the response of the human eye, which has its peak sensitivity at 5500 Å, is still fair at 4500 and 6500 Å but drops appreciably beyond. Thus we can detect all the above lines except 3727. The fact that most planetaries appear green tells us that the region of 5000 Å is the most important.

The strengths of all the lines are needed for a magnitude figure. In practice only 5007 and H beta have been measured for most nebulae; this, however, is adequate. We can count as follows: H alpha = 3 x H beta, but the eyes' sensitivity is down by about a factor of three there, so the two lines contribute equally. Add H gamma and the weaker hydrogen lines and we have about 2.5 x H beta. 4959 = one third of 5007, so count 4/3 times 5007. The N II pair can be strong, but will not contribute even an extra tenth of the light.

This is converted to a magnitude since we know the energy received by the eye from a zero magnitude star. It is a little under 10 ergs/sec/square cm of collecting area. So our magnitude becomes.

$$-2.5 \log (1.2 \times 10^5 (5/2 \times \text{H beta} + 4/3 \times 5007)$$

where the lines are measured in ergs/sec/cm^2.

We thus get the total magnitude of the nebula, irrespective of its size. To this we must add the light from the central star if the two are unresolved. The following table has been constructed from all available data. There may still be inconsistencies for objects with unusual spectra or for which the data are not reliable, but it should represent an improvement over existing tables.

Table 5. Visual Magnitudes of Planetary Nebulae.

NGC	Neb only	Neb + *	NGC	Neb only	Neb + *
650/1 (M76)	11.0	11.0	6772	13.7	13.7
1501	12.0	11.6	6778	13.0	12.8
1535	10.4	10.4	6781	11.8	11.7
2022	12.4	12.2	6790	11.3	10.1
2371/2	11.9	11.6	6804	12.6	12.1
2438	11.8	11.8	6826	9.4	8.7
2440	10.3	10.3	6833	12.8	12.7
2610	13.5	12.1	6842	13.7	13.6
3242	9.1	9.0	6853 (M27)	8.1	8.1
3587 (M97)	10.5	10.5	6857	14.1	13.5
4361	11.4	11.0	6879	13.2	13.2
6058	13.7	12.5	6884	11.6	11.6
6210	9.2	9.1	6886	12.2	12.2
6309	12.1	11.7	6891	11.1	10.6
6439	13.3	13.3	6894	12.9	12.8
6445	11.8	11.8	6905	11.8	11.7
6537	13.5	13.4	7008	11.7	11.3
6543	8.7	8.6	7009	9.1	9.0
6567	11.6	11.6	7026	11.6	11.5
6572	8.9	8.5	7048	12.8	12.8
6720 (M27)	9.6	9.6	7354	11.7	11.7
6741	12.3	12.3	7662	9.2	9.2
6751	12.6	12.0			
IC					
351	12.6	12.5	4593	11.2	10.1
418	9.6	9.5	4634	11.7	11.7
1747	12.8	12.7	4846	12.4	12.4
2003	12.3	12.2	4997	10.9	10.8
2149	11.2	10.1	5117	13.3	13.3
2165	11.3	11.3	5217	12.0	11.9
Other Nebulae					
Hu1-1	13.1	13.0	Me1-1	12.7	12.6
J 320	12.6	12.2	Hu1-2	12.8	12.8
J 900	12.2	12.1	Hb 12	12.7	12.7
Hu2-1	11.9	11.9			

3. OBSERVATION OF PLANETARY NEBULAE.

For the amateur these nebulae can afford a great deal of interest, and, as with all objects, the larger the aperture employed the greater the number that will be visible: the resolution of the brighter sources will also be greater.

Before going further into observation methods, it is as well to dwell for a short time on the problems of magnitudes and surface brightness.

LIMITING MAGNITUDES.

In Sky and Telescope for June 1973, the results of various observers' work on limiting magnitude tests were compared with computed threshold magnitudes for various apertures. The observed figures compared well with the predicted ones. To take three examples of computed limiting magnitudes, for a 6-inch aperture lim. mag. = 13.6: 8-inch = 14.2: 12-inch = 15.1.

Limiting magnitude will be greater in the zenith than at lower altitude, and with extended objects the decrease in visibility can be even more acute with decreasing altitude. Although limiting magnitude figures are given for stellar sources, and may not be thought to be valid for nebulae, it will be seen further on that certain planetaries do come within range of a telescopes' limiting magnitude. Even so, extended sources of not too great angular diameter can be observed to quite low magnitude levels, given adequate observing conditions.

MAGNITUDES, SURFACE BRIGHTNESS AND VISIBILITY.

Although the actual dimensions of planetary nebulae are very large, the appearances they display to the observer vary enormously. The difference between the apparent sizes and magnitudes of objects depends not only upon the intrinsic properties of the nebulae themselves, but also upon their distances.

The magnitude given for an extended object is the integrated stellar magnitude. For this reason magnitude figures for some nebulae can often be quite misleading, and another factor enters into our reckoning. This factor is the surface brightness (SB) of the object, of which there are two kinds, the apparent and the true.

True SB is the energy emitted per cm^2 of the surface of an object: apparent SB is the brightness per unit area, often being designated as the magnitude per square arcsecond. SB is thus a prime factor in the visibility of any extended source, and with increasing angular diameter the more difficult an object will often be to observe, even if its magnitude is quite high.

For visual observations of planetary nebulae, probably the

most notable object from the point of view of high magnitude and low SB is NGC 7293. The magnitude of this nebula is 6.5, but its angular diameter is 900 x 720 arcseconds, and unless conditions are fine, this object can fade completely into the sky. An extra disadvantege for UK observers is the low altitude that this nebula reaches, even when on the meridian, thus adding atmospheric absorption to the difficulties already outlined. These large, diffuse nebulae eventually reach such low surface brightnesses that even visual observation with the large reflectors will not show them.

As one descends the scale of angular diameters, planetaries become more apparent visually, and what might be termed 'mid-point nebulae', like NGC 3242, will be found to be easily visible.

There do exist planetaries of much fainter magnitudes and smaller apparent dimensions than not only NGC 7293, but also other nebulae that, compared with the latter, are of small diameters. These are the stellar and quasi-stellar nebulae, and although it may be thought that these show only a stellar image, such is not always the case. These nebulae are very numerous, and Figure 1 shows a plot of selected nebulae, where the increase in numbers as angular diameters and magnitudes decrease can be seen.

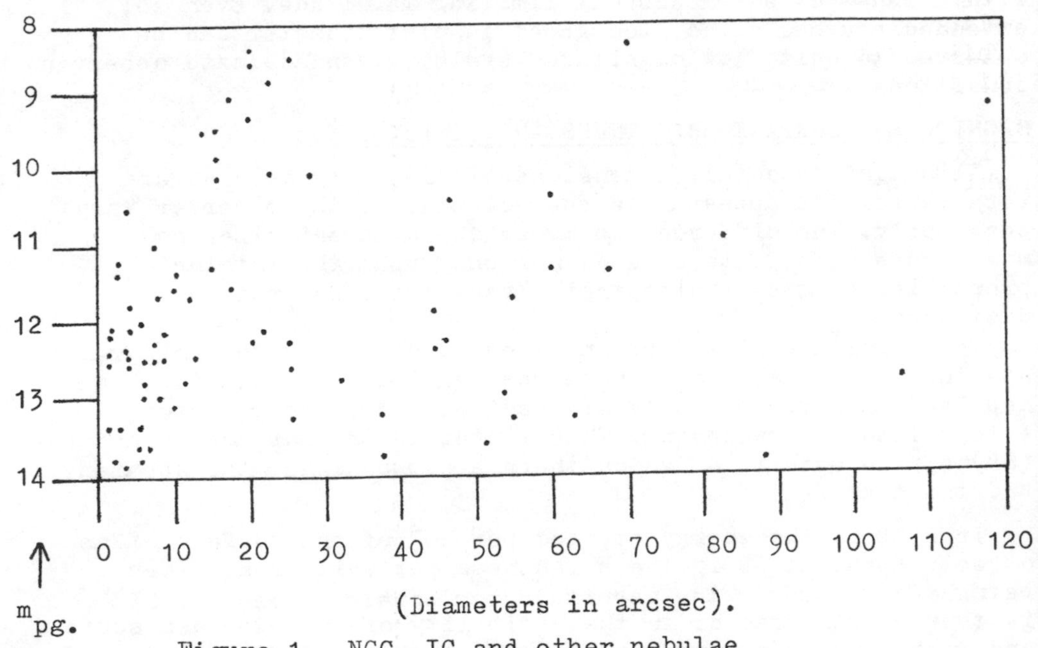

Figure 1. NGC, IC and other nebulae showing increase in numbers with decreasing parameters.

Observation of Planetary Nebulae.

Although many of these objects will appear stellar regardless of the magnification used, others will reveal circular, elliptical or even irregular shape.

The apparent dimensions of these objects range from about 2 to 10 arcseconds, so it will be appreciated that at low power they appear completely stellar. In fine seeing, when images are not enlarged by atmospheric turbulence and temperature fluctuations, and high powers can be used to advantage, those nebulae over about 4 arcseconds will often differ enough from nearby field stars of similar magnitude to enable the difference to be seen.

PRISM OR DIRECT VISION SPECTROSCOPE OBSERVATIONS.

There are certain aids to observation of the smaller nebulae which enable them to be located in the often rich star fields they inhabit. These instruments are the direct vision spectroscope and the prism.

Small, slitless spectroscopes, employing a chain of three prisms, can be used for observation of the spectral features of bright stars, and they can also be used for the detection of small planetary nebulae. The spectroscopes are not always easy to obtain, but a good substitute is the prism of 30° angle or so. This is simple to use, and the only drawback, although a slight one, is the slightly distorted field of view. This effect can be corrected to a degree by slight alteration of the prism's angle relative to the eyepiece.

If the prism is set against the eyepiece, as depicted in Figure 2, the field stars will be spread out in their spectra, while any small planetary will remain as a stellar image.

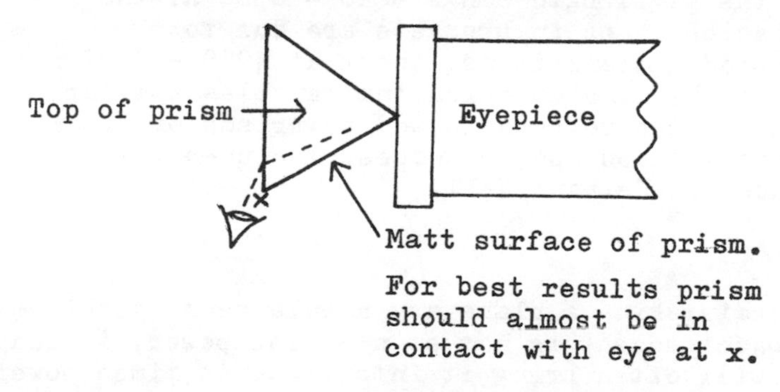

Figure 2. Top view of relative positions of prism and eyepiece. Observer should be almost at a right angle to the drawtube/eyepiece/prism extension.

Observation of Planetary Nebulae.

The reason why the image of the planetary remains stellar is to be found in its spectrum. Figure 3 shows the main spectral features of planetary nebulae, the strength of the forbidden lines of twice ionized oxygen at 4959 - 5007 Å being clear, as described in Chapter 2. These bright emission lines, being so close together under the small dispersion achieved by a prism, appear as a single emitting region. With the lack of spectral widening, the planetary thus appears as a single point-like image through the prism, the emission lines elsewhere in the spectrum being fainter, although some of these may be visible through the larger amateur telescopes.

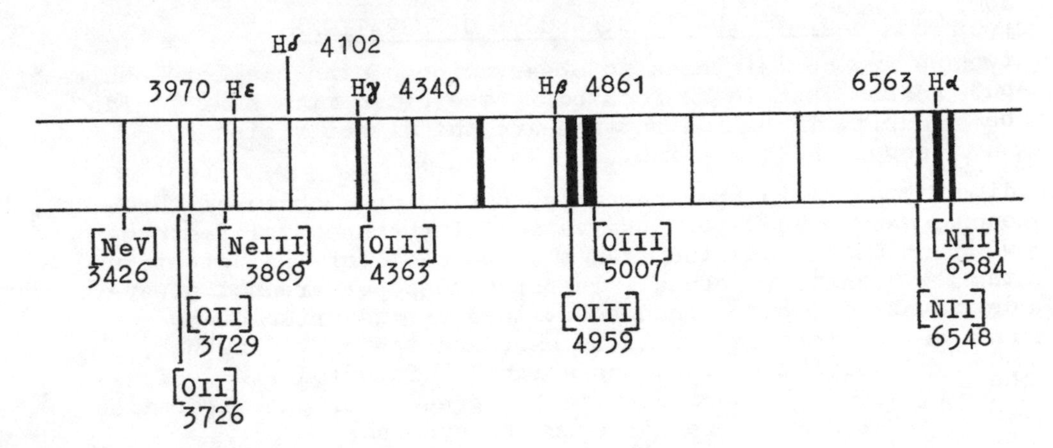

Figure 3. Main spectral features of planetary nebulae in the wavelength range 3426 - 6584 Å. The emission lines in brackets are due to forbidden transitions, those at 4959 - 5007 Å being the line emission that enables stellar planetaries to be detected by prisms or by direct-vision spectroscopes. The un-numbered lines are HeI and HeII.

There is no mistaking a planetary nebula seen by this method, and, if the object sought is not seen at low power, increase in magnification will often bring it into view. At times powers in excess of x200 are neccessary to achieve this, in which case, due to the prism, the visible field becomes even more restricted than is normally the case with the use of high powers.

When using this method on a faint planetary, it is an advantage to use some kind of head cover, particularly when the fainter sources are being observed. In the observation of all faint objects, in fact, a head cover is an asset, as, to get the most out of the object under scrutiny, only the light from the eyepiece should come into contact with the eye. The degree to which the eye is open to light from the night sky should not be underestimated.

Once the prism has been used to find the nebula required, higher powers can then be used normally to detect any shape or structure. With objects of the smallest angular diameters no detail is seen, but in others basic shape can be observed. Some of these nebulae are oval in shape, e.g., IC 5217; irregular, Hu1-2, or decidedly elliptical, M1-7. The first two of these three nebulae are quite easy objects to detect, but the latter is of very low surface brightness, and can be classed as quite a severe test for moderate apertures.

One can get along perfectly well without a prism or direct vision spectroscope, plenty of nebulae being large enough to need no such equipment. Even so, the use of either will greatly extend the number of observable nebulae, and the fact that many show a certain, if limited, amount of detail will make the observations suitably interesting.

At this point it is worth mentioning two cases that we have come across where the respective nebulae, although of small angular sizes, were detectable without the use of a prism, one due to its faintness, the other to its brightness. For the first of these, M2-2, the prism was found to be a disadvantage. With angular dimensions of 12 x 11 arcseconds, M2-2 is not much different from other nebulae that are easy prism objects. The disadvantage of the prism is related to the very low surface brightness of this object, and the final successful observation was made normally, a point worth bearing in mind in cases where attempts with a prism are fruitless, and direct observation may well prove to be the answer.

The second object is IC 2165, which we have identified by appearance alone and then confirmed with the prism, and very small nebulae of sufficient brightness should differ in the following way from nearby stars. Firstly, IC 2165 is not purely stellar; its photographic dimensions are 9 x 7 arcseconds, but this must be reduced somewhat to arrive at a meaningful visual size. Therefore, at low power, this nebula will appear stellar. How, then, is it possible for it to be identified without prismatic aid?

The answer appears to be that, small as it is, this nebula is a disk. Observation showed it to have rather a 'flat' look, i.e., lacking the scintillation of nearby stars of abou equal magnitude, and we put this down to the disk presence. A good analogy is the difference in naked-eye observation of a planet and star of equal magnitude (Saturn and Capella at times) where the disk of the planet considerably reduces the scintillation apparent in the appearance of the star, a point source.

NEBULAE OF MEDIUM ANGULAR DIAMETERS.

With moderate apertures, differences in nebular distribution can be observed in many planetaries, from fairly regular ring structure, (NGC 6720 - M57) to broken ring structures (NGC 7008) and rather irregular nebulosity (NGC 2440).

It is with many of these objects that it is always worth trying the use of high magnification, as this can bring out structural differences to a good degree. At this point we must consider certain features of some of these nebulae and how these relate to their visual appearance.

Ring Nebulae.

The prototype of this class of object is M57, and two main structural details inherent in most ring nebulae can be seen in M57 with ease. These are a) the excess brightness of the nebulosity forming the minor axes of the ring, and b) the fainter material comprising the ends of the major axes, the latter also being less well defined.

With other similar but lower surface brightness objects, the major axis nebulosity is much fainter, and the apparent major axis is actually the minor axis. A nebula of this type is NGC 6778, and in Figure 4 we show, in diagrammatic form, the distribution of material in this nebula with M57 as a comparison object.

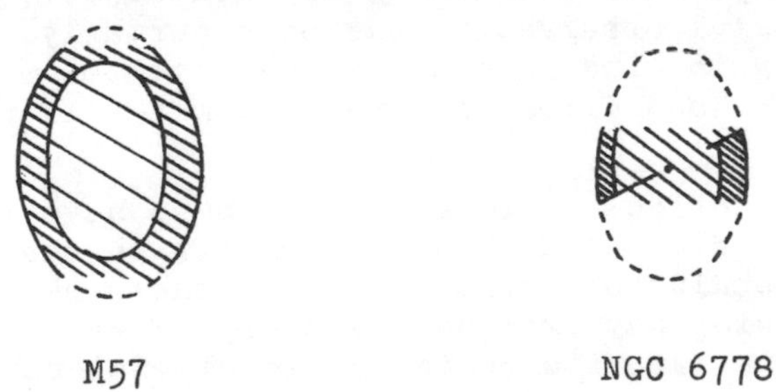

M57 NGC 6778

Figure 4. Nebular distribution in M57 and NGC 6778

(Not to scale).

Observation of Planetary Nebulae.

As well as medium sized nebulae like M57, many planetaries of much smaller angular dimensions also show partial or complete ring structure. NGC 7662 reveals indications of annularity in a moderate telescope, although high magnification is required in order to break the overall high luminosity. Another ring nebula is IC 418, and according to observations on hand, only 16½ and 18-inch telescopes have managed to resolve the annularity. Even IC 2165, referred to on page 17, is a ring nebula, but it is doubtful whether the majority of amateur telescopes could resolve this.

Other fainter but larger examples of ring nebulae are NGC 6781 and NGC 7048. In both these cases resolution of detail is not an easy matter with telescopes of about 8-inches and below.

OTHER NEBULAE.

Two planetary nebulae of interest are NGC 2346 and NGC 2440; in these the visual appearance is more amorphous, and the latter is classed by Gurzadyan as diffuse.

NGC 2346 has been called the 'Hour Glass Nebula' and the waisted appearance of this object is well defined on long exposure photographs. Visually, this nebula resembles not so much a planetary as an H II region; amorphous, fairly faint, but nevertheless easy enough to observe. With the larger telescopes employed by amateurs some structure can be seen, (cf. Malcolm Thomson's observation in the catalogue), but still nothing relating to the actual outline.

The second nebula, NGC 2440, while basically irregular in shape, is a quite high surface brightness object in which a certain degree of detail can be made out. In its basic shape and luminosity distribution it compares well visually with its photographic appearance.

Other planetaries, e.g., NGC 1535, M27, show structure of some kind also. From the results set out in the catalogue further on, it can be seen that the percentage of planetaries which do show detail, however minimal, is quite high.

COLOUR IN NEBULAE.

The spectral region 4500 to 6500 Å is that to which the eye is mainly receptive, the sensitivity in colour being greatest in the green and yellow. Colour in nebulae of large angular size is not apparent, but for the smaller objects it is often clear, and many observers have noted the green, green-blue or blue colour of these nebulae.

Medium bright objects such as Hb 12, which is strikingly green in large apertures, and compact, bright nebulae like

NGC 6572 and NGC 7662 are the objects that will display colour.

It must be stressed that many factors are in operation here; the colour sensitivity of the observer's eyesight, the aperture used, the quality of transparency and the zenith distance of the object all contributing.

NUCLEI OF NEBULAE.

Nuclei of planetary nebulae appear in a variety of guises. In some cases they visually dominate the associated nebula, as with NGC 1514. The nucleus of A 36, a nebula totally beyond visual observation, is very bright, (11.5 mag).

In the case of NGC 6543 the 9.6 magnitude nucleus is well enveloped in dense nebulosity, and moderate apertures appear to find difficulty in registering this star. Use of a prism on this object will show the spectrum of the nucleus projecting through the nebulosity.

NGC 40 is another nebula whose nucleus can be of interest to prism or spectroscope users. This star is of 11.6 visual magnitude, and is very rich in bright emission lines (class WC). At high power the prism or spectroscope should show the star as displaying a beaded appearance due to this line emission.

OBSERVATIONAL CHECKS.

If observed structural details of an object are considered to be doubtful, then further checks are neccessary. These must be made in (as nearly as can be ascertained) equivalent or superior conditions. A method of achieving this is to use a selected object upon which to check transparency. With the sky being so variable, there may be some time lag between the two observations, but this is of no real account, as quality, not quantity of results should be the aim.

FINDING FAINT OBJECTS.

Many planetaries of all sizes can often pose problems of finding. In cases like this the SAO Star Atlas is beneficial, as the position of the nebula relative to field stars can be checked. Even if the reference stars are some distance from the object sought the telescope can be offset from them. The SAO Atlas is also useful for those whose telescopes are not fitted with setting circles, as the configurations of field stars near the nebula sought can be used as guides.

4. RECORDING OBSERVATIONS.

Once an object has been located, the business of recording the observation should be carried out in a systematic manner. If the observation is to be a purely descriptive one, then the procedure should be along the following lines.

(1) INITIAL APPEARANCE.

The appearance of the object should be recorded first at the lowest power it is visible at, and any characteristics double checked before being commited to paper. The relative position to close field stars should be noted, as well as determinations of the magnitudes of key field stars.

(2) APPEARANCE AT HIGHER MAGNIFICATIONS.

The subsequent observations should attempt to determine any structural features not visible at low power. Often these can be considerable, and the observation should not be concluded until it is felt that all available information has been recorded.

If the object displays different degrees of brightness, then these should be recorded fully and details of their position within the object noted. Also neccessary are determinations of the angular dimensions of these sections and their relative position to other parts of the object, plus whether or not the details become clearer with higher magnification.

The whole range of magnifications that conditions will allow should be used, either to resolve detail more fully, or to attempt to discern any other structural aspects that were not visible at lower powers.

Doubt about any object or detail seen should be recorded as such, and, where neccessary, follow-up observations made.

(3) FIELD ORIENTATION.

Finally the field orientation should be checked, as should the PA of the object and the PA of any field stars near to or superimposed upon the object.

(4) DRAWINGS.

These can act as supplements to written work, and should be as accurate as possible. They should be made up in circles of suitable diameter for the field of view of the eyepiece in use. Ideal sizes are 4" diam. circle for a LP field; 3" for MP and 2" for HP.

Estimates of field star magnitudes can be placed alongside the respective images. If it is felt that observed structure requires a larger scale drawing, then this can be made to act as a supplement to the main drawing. In all drawings the field diameter in arcminutes should be indicated.

Recording Observations.

(5) OBSERVING CONDITIONS.

Full details on seeing and transparency are essential; the scale of seeing drawn up by Antoniadi, as described in the Double Star Handbook, should be used.

I. Perfect seeing without a quiver.
II. Slight undulations with moments of calm lasting several seconds.
III. Moderate seeing with large tremors.
IV. Poor seeing with constant troublesome undulations.
V. Very poor seeing.

A similar scale for transparency should also be used.

(6) ACCURACY OF RESULTS.

Within the limits of visual work the greatest accuracy possible should be the aim. Naturally, with the different approaches and qualities of observers, the word 'accuracy' is a variable quantity. For all that, the use of vague terms should not be part of the observational procedure. Vague terms can be those such as 'lucid star', 'bright star', 'near' or 'distant'. What is bright in one aperture is not so in a smaller one; near can be anything from 4 to 40 arcseconds; distant anything from 4 to 40 arcminutes.

Sizes of objects or parts of objects should be determined from reference points in the field or from the whole or half field diameter.

Thoroughness in observation means greater interest in the observational process. Visual observers are not working at the levels of professionals, but, within their individual limits, they should certainly try to achieve the greatest possible accuracy.

PART TWO : GASEOUS NEBULAE.

INTRODUCTION.

The topic of this Webb Society Handbook is nebulae, and it is appropriate to define at the outset what we mean by this term. The word comes to us directly from the Latin, where nebula (pl. nebulae) means mist or cloud. In the astronomical sense this is taken by most popular dictionaries to imply an unresolved group of stars, or a cloud of glowing gas; something which looks like a small patch of cloud. This, however, is an unsatisfactory definition. The simple fact that a compact cluster of stars cannot be resolved by the naked eye or a given telescope does not imbue it with any useful property which a resolved group does not share. So we have adopted a more precise, more critical definition. To be included in this section of the Handbook the nebulae must be gaseous.

In practice very few gaseous nebulae would be visible to us without associated stars. Except for supernova remnants, all bright gaseous nebulae are illuminated by stars, often in clusters, which may be more obviously visible than the nebula itself. Everyone knows the Pleiades as a cluster, but few have seen the complex of gaseous nebulae which surround the Pleiad stars.

An important constituent of gaseous nebulae is dust. In the average gaseous nebula there is about 10 grams of dust for every kilogram of gas, and this would normally be spread throughout a volume of one million cubic kilometres. The dust exists as tiny grains of sand and graphite, about a millionth of a centimetre in diameter and typically ten cm apart. This sounds very tenuous, but when accumulated into the vast interstellar clouds the effect of the dust becomes marked.

It is the presence of dust which makes two types of gaseous nebulae visible. One type, the obvious manifestation of dust, is the dark nebulae seen in silhouette against a brighter backdrop. Perhaps the best-known of these are the 'Coal Sack' and the 'Horsehead' nebula in Orion. Reflection nebulae, such as those round the Pleiades, are made visible simply by reflected light from nearby stars. Tenuous gas alone does not reflect light efficiently; it is the tiny dust particles which make the nebula visible.

In order that the gas itself shine it must be excited by extreme ultraviolet radiation. Sufficiently hot stars (in practice O or early B stars) can provide this ultraviolet energy. Good examples of suitable stars are members of the Trapezium (theta 1 Orionis). These stars provide sufficient ultraviolet radiation to excite the whole of M42 and many other bright nebulae in Orion besides. The hotter the star, the more ultraviolet radiation it emits and the brighter the nebula, provided only that there is sufficient gas.

Emission nebulae come in three main types. H II regions are irregular clouds where young hot stars have recently formed to illuminate them. Planetary nebulae are compact, often circular, nebulae in a late stage of stellar evolution. Supernova remnants are the gaseous remains of stars which have blown off their outer layers in a supernova explosion.

Each of these nebulae is described in a separate chapter of this Handbook, (planetary nebulae being separately covered in Part I). In addition Chapter 10 describes some extragalactic nebulae, using the definition of nebulae given above. It is not often realized that extragalactic nebulae - as opposed to galaxies composed predominantly of stars - can be observed with quite small telescopes. The brightest such object is 30 Doradus, the cluster and H II complex at the heart of the Large Magellanic Cloud, and this is visible to the naked eye.

Apart from the few extragalactic objects, gaseous nebulae lie in the spiral arms of our galaxy and hence close to the Milky Way. They are seasonal, late summer to midwinter being the appropriate time of year in both hemispheres.

HISTORICAL REVIEW.

Gaseous nebulae, being generally both faint and diffuse, escaped detection before the invention of the telescope. Even then they were slow to reveal themselves. As might be expected the first such object to claim attention was the huge complex of glowing gas and dust comprising what was previously thought to have been a single star - theta Orionis. The Great Nebula in Orion was first discovered to be 'nebulous' by Nicholas Pieresc - a friend of Galileo - in 1610, the Jesuit astronomer Cysatus making an independent discovery some eight years later. Credit for the first detailed description and drawing of this famous nebula, however, belongs to Christian Huyghens, who in 1656 also noted three of the involved stars which later became known as the Trapezium, (the fourth star was first discovered by Picard in 1673). A detached portion of the Orion Nebula was observed by de Mairan some time before 1713, and Messier included this as a separate object - M43 - in his catalogue of 1771.

The 'Omega' nebula, M17, in Sagittarius, was the next to be discovered by de Chesaux in 1746, and this was followed by Le Gentil's recognition in 1749 that the star cluster in the bow of Sagittarius (first found by Flamsteed in about 1680) also contained nebulosity: this object, catalogued by Messier as M8, later became known as the 'Lagoon' nebula.

The Abbe Lacaille, during his observations at the Cape of Good Hope in 1751-3, attempted to categorise the nebulous objects he found into three classes, viz., 1) Nebulae without stars,

2) Nebulous stars, 3) stars with nebulosity. Later observations have shown that Lacaille's divisions had little relevance to actual physical classification, but among his discoveries we may note NGC 2070 in 30 Doradus and NGC 3372, in which the one-time spectacular variable eta Carinae is involved.

Messier himself was responsible for the discovery of M20, but he saw it only as a star cluster, the gaseous nebulosity being first described as 'Trifid' by John Herschel in about 1830. Another bright portion of gaseous nebulosity in Orion, M78, was discovered by Méchain in 1780. Thereafter, with the aid of vastly improved telescopes in the hands of Wm. Herschel, discoveries followed thick and fast, but even this great pioneer of astronomy was never able to satisfy himself completely whether or not all nebulae were ultimately resolvable into stars. This matter was eventually settled by Wm. Huggins in 1864, when with the aid of a spectroscope he showed that the planetary nebula NGC 6543 in Draco gave a spectrum similar to that of a 'luminous gas'. On the basis of these observations, Huggins suggested the division of nebulae into two classes, 1) 'green' nebulae (with a gaseous spectrum) and 2) 'White' nebulae (with star-like spectra). This division effectively separated gaseous and planetary nebulae from the unresolvable extragalactic nebulae (galaxies) and it is at this point that a more meaningful analysis began.

5. CATALOGUES OF GASEOUS NEBULAE.

Nebulae do not feature prominently in most catalogues and star atlases which are, instead, usually filled with star clusters and galaxies. Out of the full list of 109 Messier objects there are only seven gaseous nebulae, made up of five H II regions, one reflection nebula and one supernova remnant.

The assiduous observer will find a few more examples buried in the charts of Norton's Star Atlas, but if he relies solely on the "interesting objects" lists he will find only a further three objects for his study, one a reflection nebula and two southern hemisphere H II regions.

With these objects satisfactorily observed, the deep-sky specialist will probably turn to Atlas Coeli for further reference. The polychrome charts of Coeli are liberally spread with vivid green patches, representing bright nebulae, and dusky areas indicating dark nebulae. In the accompanying catalogue Becvar lists most of the objects drawn on his star charts - 240 "bright diffuse nebulae". Since this is the most complete list which normally is available to or is bought by amateurs, we will analyze it critically.

The catalogue of bright diffuse nebulae is of very little value. First, several of the entries refer to non-existant or mis-classified nebulae. These errors are detailed in Table 1.

Table 1. Errors in Atlas Coeli Gaseous Nebulae.

Object	RA	Dec	Remarks
IC 11	00 17.8	+56 19	Not real.
	48.1	+57 50	Cluster; no nebulosity.
IC 155	01 44.0	+59 32	Not real; possibly a misplaced representation of Sharpless 188, a faint emission loop S.p. Chi Cas.
IC 1851	02 48.0	+58 06	Probably not real.
NGC 1465	03 50.5	+32 21	Galaxy.
	04 03.3	+27 29	41 Tau; no nebula.
	18.8	+28 20	Not real.
	38.0	+27 00	IC 2088 complex of nebulae; not real.
	05 26.0	+12 31	Not real.
IC 419	27.8	+30 07	S. of T Aur; not real.
	36.0	+13 35	Not real.
NGC 2045	42.2	+12 52	Not real.
	06 10.0	+17 05	Not real.
	20.6	+04 57	N. of 8 Mon; not real.
NGC 6360	17 22.2	-29 57	Not real.
IC 4657	29.0	-17 29	Probably not real.
IC 4659	31.3	-17 54	Probably not real.
IC 4678	18 04.9	-23 53	Faint extension to M8; not separate.

Catalogues of Gaseous Nebulae.

IC	4683	18 05.2	-23 25	Not real.
IC	4681	05.6	-26 15	Not real.
NGC	6820	19 40.4	+22 58	Single nebula only; nebulae marked to NE and NW not real.
IC	1307	40.7	+27 23	Not real.
		20 22.2	+38 21	M29 and three nebulae S.p. it; none real.
		52.7	+46 32	N.f. 55 Cyg; not real.
IC	1369	21 10.4	+47 33	Cluster; no nebulosity.
		10.7	+59 47	Not real.
IC	1400	42.5	+52 43	Not real.
NGC	7748	23 42.6	+69 29	Not real.
IC	5366	55.1	+52 31	Not real.

(Positions in the above Table for 1950.0).

Secondly the catalogue gives no indication whatsoever of the surface brightness of the nebulae and hence their feasibility in amateur telescopes. Thirdly a very large proportion of the nebulae are too faint to be recorded with the size of telescopes amateurs normally employ, and many defy visibility through the largest telescopes in the world. Even the photographic emulsion which is superior to the human eye at recording faint nebulosity, will not make much impression on many of these objects unless large telescopes are employed. The amateur is therefore encouraged to waste time looking for objects he stands no chance of finding. Finally there are a significant number of moderately bright nebulae not included in the Becvar catalogue.

It is to remedy these shortcomings, and to present a catalogue of nebulae which will be of use to the interested amateur, that this Handbook has been compiled. We have aimed at a reasonably complete catalogue of the objects that amateurs can hope to observe, but we are aware that this, like its predecessors, will contain omissions and will suffer other shortcomings.

It is valuable here to mention briefly the catalogues of nebulae used by professionals, all of which are available in good astronomical libraries such as that of the Royal Astronomical Society.

Bright Diffuse Nebulae.

Catalogues of these objects have been compiled from studies of the Palomar Observatory Sky Survey plates, and list only the position, approximate size and a crude brightness criterion. Unfortunately for amateurs, any nebula bright enough to be visible in a small telescope is grossly overexposed on PSS plates, so no useful brightness classification exists for them. The most complete list is given by Beverley Lynds (1965, Astrophys. Journal

Supplement, 12, 163), but for H II regions reference is usually made to Sharpless (1959, Astrophys. Journal Supplement, 4, 257); most faint H II regions are known by their Sharpless number. A corresponding catalogue for the southern hemisphere is Rodgers, Campbell and Whiteoak (1960, Monthly Notices R.A.S., 121, 103).

Dark Nebulae.

Various photographic catalogues of dark nebulae have been compiled, e.g., by Barnard and Bok. The most complete, based again on Palomar Sky Survey Prints, is that of Lynds (1962, Astrophys. Journal Supplement, 7, 1).

6. DARK NEBULAE.

Any nebula which contains dust will partially or completely obscure whatever lies behind it in just the same way that a suspension of small particles in air - smoke - can be opaque. If no stars illuminate the outer face of the nebula it will appear dark, and its presence can therefore be detected optically by its obscuration of whatever lies behind it. If the background is a bright nebula or particularly dense region of stars it will be seen as a "hole" in the sky. If not, it will pass unnoticed. At present nobody knows how many dark nebulae lurk in the sky with no background to show them up. There is, however, evidence that there are extensive thin clouds over a large area towards the south galactic pole, for in this direction significantly fewer quasars can be detected optically even though many radio sources exist.

The best equipment for studying the dark nebulae is the naked eye. Go out on a dark summer night and look at the Milky Way in Cygnus and Aquila. The Great Rift which splits it into two is one example of a dark nebula. If you study it carefully you will find it to be resolved into a number of large and small blobs joined together into a long, sausage-shaped dusky band. Look at it through a small telescope and it becomes very hard to see. On deep photographs the Great Rift vanishes because it is not really very dark.

Much better known to southern hemisphere observers is the 'Coal Sack' which Crux sits astride. This single nebula somehow contrives to look darker than any other patch of the entire sky, yet it too begins to disappear on photographs. Bart Bok invented a dark constellation, the 'Emu' (or 'Ostrich', according to one's persuasion), the 'Coal Sack' is its head, and there is even a dark beak projecting from one end and a 6th mag star to form the eye. A narrow neck of material stretches between alpha and beta Centauri; the body swells out in Sco-Sgr, and the thin dark legs climb into Aquila.

The 'Emu' and many other gaps in the Milky Way are the dust clouds which accompany the spiral arms of our galaxy, just as they do in other edge-on spirals like the 'Sombrero Hat' (M104). Indeed, these dust lanes are one of the strongest pieces of evidence that our galaxy really is spiral.

Not all the dark nebulae are so big, however. Bok himself found and listed many tiny, oval dark nebulae which have become known as Bok globules. Some are up to several minutes of arc across and occult star fields near the galactic centre; others are only a few seconds of arc in diameter and are seen as dark spots on photographs of H II regions such as M16.

But it is in the Orion Nebula that the telescope owner will

find one of the most impressive dark nebulae. Known as the 'Fish's Mouth', it intrudes into the nebula north of the Trapezium. This is not a gap in the nebula, but a finger of gas and dust curling around in front of M42.

It is worth describing M42 in a little more detail. Most observers think of it as an H II region, a bright nebula. But this is really only a small part of the real Orion Nebula complex.

The key to our understanding of the Orion Nebula complex is the discovery that molecules exist in space and radiate at specific wavelengths in the centimetre and millimetre region of the spectrum. The main molecules of interest are carbon monoxide and formaldehyde. Both occur in large dense clouds of gas and dust far from any hot stars, i.e., in dark nebulae. One of the most intense CO and H_2CO clouds is in Orion. It has two intensity peaks, one about 1' arc north of the Trapezium and another almost 15' north of that. The entire molecular cloud covers an area greater than that of the moon, includes a number of infrared sources which may be stars in the making, and must contain enough material to form 100,000 stars like the sun.

A small group of very young, hot stars has formed in the Orion Nebula very close to our edge of it. They are sufficiently bright to blow away the dust grains that obscure them by radiation pressure, thus creating an expanding clear bubble in the nebula. Probably a few aeons ago the bubble finally burst through and the stars became visible; we now call them the Trapezium. Another star, right at the outer edge, illuminates a small patch of this giant nebula; this is M43. The 'Fish's Mouth' is the last remnant of the dark nebula which once obscured the Trapezium stars from our view. It too is being irrevocably eroded by their radiation, and in a further few aeons will have vanished.

There may be other Trapezium-like stars or clusters burning their way out of the Orion Nebula. Some, on other sides, may have succeeded. If not, there will exist directions from which the nebula is quite invisible. Other creatures living in other parts of our galaxy, if any exist, may be quite unaware of the Orion Nebula.

Finally, to put things into perspective, we can measure the darkness of a dark nebula by the number of stellar magnitudes of extinction it imposes on light travelling through it. The Great Rift dims starlight by two or three magnitudes. The Orion Nebula produces extinction of about 50 magnitudes. It would in theory be possible to hide the most luminous quasar in the middle of the Orion Nebula and not see it.

7. REFLECTION NEBULAE.

It surprises nobody to be told that reflection nebulae shine by reflected starlight. Many do not realise, however, that it is tiny particles of dust only a few hundred Angstroms in diameter which do the reflecting. Gas as tenuous as that in nebulae does not reflect light.

Reflection nebulae, then, are curtains of gas and dust lying near stars. Stars in association with gas are usually young, having recently formed from the gas and not yet had time to move away or disperse the material. Many T Tauri stars, canonically very young, therefore have reflection nebulae, and Hind's variable nebula around T Tauri itself is a good example. T Tauri stars are, however, very faint, and their nebulae are fainter still. Thus few of them have found their way into this part of the Handbook. Instead, the illuminating stars are quite bright.

Before he turned his attention to extragalactic objects, Edwin Hubble spent some time investigating reflection nebulae. He demonstrated their nature most clearly by showing that the size of the nebula - i.e., the distance from the star at which it could just be detected - depended upon the magnitude of the star according to the inverse square law. Hubble's relation does not hold for emission nebulae.

Dust grains, like all small particles (including the aerosols in the Earth's atmosphere) reflect blue light better than red. Reflection nebulae are therefore bluer than the stars they reflect, as the sky is bluer than the sun. Since the stars which produce the brightest reflection nebulae are hot and therefore themselves quite blue, most of the reflected light is in the ultraviolet where the human eye is insensitive. The combination of the extreme blueness and the inevitable proximity to a bright star prescribed by Hubble's relation makes reflection nebulae particularly difficult to observe. The most easily observed nebulae are those in which there is above average dust (and gas) content increasing the brightness close to the star, or where the star is dimmed by intervening dust more than the nebula.

It is a common thing to assume isotropy in astronomy - that things look much the same in any direction. Thus if stars are surrounded by gas and dust nebulae, these are assumed to be equally thick and extensive in all directions. Much of the time when the nebula cannot be resolved this is a good approximation, but a glance at the sky shows that nature sometimes favours flattened disks, as is the case with the Milky Way system. Disks are present on a small scale too, in the dust clouds around some stars. The solar system is evidence that the sun once had a disk-shaped dust nebula.

In the early stages of disk formation there will be a concentration of material near the equatorial plane, but still considerable amounts over the poles. Seen edge-on the star will be dimmed or completely obscured by the equatorial dust, but light escaping towards the poles will be reflected by the dust there.

Such nebulae exist and are called "cometary". In appearance they are characterized by unusually faint illuminating stars with fan-shaped reflection nebulae spreading out like lighthouse beams in two opposite directions. In some cases only one polar fan exists; if the observer lies slightly out of the equatorial plane, the dust may also obscure one of the fans. This is the situation in M1-92, the 'Footprint' nebula in Cygnus, where one lobe (the sole) is much brighter than the other (the heel). M1-92 is probably the brightest reflection nebula in the sky, but its illuminating star is totally obscured by dust, except in the extreme red, where its light feebly penetrates.

8. H II REGIONS.

In contrast to reflection nebulae, some gas is made luminous by the presence of hot stars. In one sense such objects are like the reflection nebulae just discussed, for they require a nearby star to render them visible. The important difference, however, is that the starlight is reprocessed by the gas rather than scattered by the dust. H II regions and the other emission nebulae described later radiate in proportion to the amount of gas rather then the number of dust grains. To understand the mechanism atomic physics is involved, and the process is only sketchily covered in what follows.

It is radiation shortward of 911.5 Å, i.e., the extreme ultraviolet radiation, which excites the gas. Any photon of this wavelength or shorter is capable of dislodging the single electron which orbits the proton in hydrogen, i.e., to ionize the atom. An atom which has lost an electron is designated by the Roman numeral II, hence the term H II region. In an emission region like M42 near the Trapezium, virtually every hydrogen atom is ionized, and the gas comprises a sea of protons and electrons dashing around at high velocities.

Every so often an electron and proton meet, and the former is captured by the latter. However, the electron does not normally drop neatly back into place; instead it cascades down through a series of energy levels (electron shells or allowed orbits) giving off energy in discrete puffs. Each puff is a single photon with a wavelength characteristic of hydrogen, and in the visible region the relevant wavelengths are known as the Balmer series. In order of decreasing strength the Balmer lines appear at: 6563, 4861, 4340, 4102, 3970, 3888 .... Å. The series is endless, but all the lines merge at the Balmer limit at 3646 Å (it is no accident that 3646 = 4 x 911.5)

What the gas has achieved is the conversion of unseen ultraviolet radiation of the star into visible radiation at a number of discrete wavelengths. The hotter the star, then the greater the proportion of its energy emitted shortward of 911.5 Å, hence the more energy pumped into the Balmer lines, and the brighter the nebula. Very hot stars therefore produce very bright nebulae, thousands of times brighter than the reflection nebulae would be. For this reason emission nebulae are much easier to observe than reflection nebulae.

Helium behaves in the same way as hydrogen, (although it requires energy even further into the ultraviolet) and contributes a few more lines to the visible light; the strongest of these are at 5876 and 3888 Å. In addition, however, there are bright lines which denied identification for many years and were once thought to belong to an element called nebulium. The explanation of these features, termed, as we have seen in the introduction to Part One, 'forbidden lines', is considerably more complicated, and will not be dealt with here.

Like reflection nebulae, H II regions are associated with star formation and youth. Indeed they differ from dark and reflection nebulae only in that particularly hot stars have formed in them. Being young, all these objects lie in the spiral arms of our galaxy. However, their distribution in the arms is quite patchy. The concentration in Orion represents part of the spiral arm which passes just outside the sun at 1500 light years distance. The H II regions of Sgr lie in the inner arm which curls round between us and the centre of the galaxy. They are bigger than M42 but at a greater distance.

H II regions are strong radio sources. The electrons and protons sometimes interact without combining to give off emission lines. When this occurs the energy is radiated by a mechanism known by the German name Bremsstrahlung, or free-free radiation. Bremsstrahlung is mostly emitted in the radio region. It should not be confused with the 21 cm line emitted by neutral hydrogen; instead it has a continuous spectrum. Radio maps of the galaxy show numerous H II regions which are optically invisible because of intervening dust. Many lie in Sco-Nor, the body of Bok's dark 'Emu'.

The gas content of H II regions varies. It is measured by the number density of electrons, N_e, which can be inferred from the strengths of some of the forbidden lines. Most nebulae contain between 100 and 10,000 electrons (and hence protons) per cubic centimetre. By comparison one cubic inch of the air we breathe contains more than 10^{19} atoms.

Recently some H II regions have been found to have much higher densities. These are always small, have N_e values of about a million, and are known as compact H II regions. Exactly why they should have collapsed to such high densities before stars formed in them is not clear, but they must be extremely young else the stars would have dispersed them to lower density.

The brightest H II region is often claimed to be NGC 7027, the irregular nebula in Cygnus discovered by Webb in 1879. There is some disagreement whether to class NGC 7027 as a compact H II region or a planetary nebula. It has some characteristics of both and is perhaps unique because of its high electron density, very hot central star and enormous quantity of dust both inside and immediately outside the nebula. In this Handbook we have included NGC 7027 among the gaseous nebulae.

9. OTHER EMISSION NEBULAE.

All the nebulae described so far have one feature in common: the presence of a star to cause the gas to glow. This, however, is not a neccessary condition.

Consider the example of meteors. These tiny grains of sand collide with the gas of the Earth's atmosphere at high velocity. As they travel through it they are slowed down by frictional drag. The kinetic energy they lose is converted partly into heat, which melts and ablates the meteoroid, but partly also into ionizing the atmospheric gas. This gas gives off molecular emission lines, and it is these which produce the visible train of the meteor. No star is involved.

Even in the vastness of interstellar nebulae kinetic ionization of gas is possible. This happens when one body of gas is forced to move through another, particularly if the velocity exceeds that of sound (about 10 km/sec in the interstellar environment). When can this happen? The most obvious example is when a star explodes.

Novae have emission-line spectra from a few days past maximum for as long as they can be observed. In part this is because of the hot star which remains from the outburst and which provides ultraviolet radiation. In part, too, the gas is collisionally ionized. The exact balance between ultraviolet and collisional ionization is not yet known with confidence.

Several novae have produced nebulae large enough to be resolved. Three of these are marked in Atlas Coeli, and may just be possible for observers with large apertures. These nebulae are:

		1950		1975	
Nova Per	1901	03 28.0	+43 44	03 29.7	+43 49
T Aur	1891	05 28.8	+30 24	05 30.4	+30 25
DQ Her	1934	18 06.0	+45 50	18 06.7	+45 50

To seek unambiguous examples of collisionally ionized nebulae we must turn to supernovae. Here we can be sure of our ground because the remnant star of a supernova explosion very quickly becomes incapable of exciting gas to emission. But the massive injection of kinetic energy carries the gas from the star outwards through the sparse interstellar medium with enough force to keep it glowing for millions of years. In later life the expanding bubble of collisionally ionized gas produces an object very like a planetary nebula in its appearance, but with rather a different spectrum. The tiny remnant star of the explosion will by then have ceased to shine and will, in all probability, have been ejected by the explosion and be far away.

The large circular complex of nebulae in the S.f. corner of Cygnus, known as the Cygnus 'Loop', is the largest and brightest collisionally ionized nebula in the sky, being now a delicate filigree of emission nebulosity wreathed into a circle twice the diameter of the moon. It incorporates the NGC nebulae 6960, 6979, 6992 and 6995. Most of these five objects - the brighter parts of the rim - can be seen in amateur-sized telescopes.

Next after the Cygnus 'Loop' the brightest collisionally ionized nebula is the supernova remnant IC 443, this, however, is rather faint.

Only in 1976 was it demonstrated by Mike Dopita of Mt. Stromlo Observatory that there are other collisionally ionized nebulae in the sky. In the mid 1950's George Herbig and Guillermo Haro independently discovered some very faint blotchy emission nebulae in Orion. These Herbig-Haro objects, and others found elsewhere since, are characterized by their emission-line spectra and lack of a visible star. Thanks to the work of the Stroms at Kitt Peak and to Dopita we now know that there is an exciting star, hidden by dust but detectable in the infrared. This star has, for some reason, a very large stellar wind, and the gas driven out by this wind impinges on stationary material in the dark nebula which obscures the star, producing the collisionally ionized nebula we see. Sadly, no Herbig-Haro object is bright enough to be seen visually in any telescope.

There remains one nebula which does not quite fit into any of the categories enumerated in the previous pages. It is unique. It has been the subject of television programmes; it radiates at radio, infrared, optical, ultraviolet, X-ray and gamma-ray wavelengths; its discovery prompted the first significant catalogue of nebulae and clusters; its central star flashes on and off sixty times a second; it was once visible in daylight. It is, of course, NGC 1952, Messier 1, the 'Crab' nebula. It is easily visible in small telescopes, and is an object that everyone should observe to complete his education.

True M1 is a supernova remnant and could therefore have been introduced earlier in this chapter, but description of it was deferred because of its unique properties. In all other supernova remnants the emitting gas forms annular veils of nebulosity about a dark centre. In M1 the emission filaments are quite faint and are overshadowed by a dull yellowish light coming from the gas within. This gas has a continuous spectrum, like that of a reflection nebula, but for a completely different reason: it emits synchrotron radiation.

The central star of the Crab Nebula is an enigmatic object.

Other Emission Nebulae.

It is the very dense core of the original supernova, so dense that matter is crushed by its own gravitational force to densities millions of times higher than can be generated on Earth. The atoms themselves have been crushed and are no longer recognisable. This neutron star, as it is called, rotates 30 times a second, and as it does so it sends out two beams of radiation like those of a lighthouse. Unlike a lighthouse, the neutron star sprays out both radiation at every wavelength known to man and electrons at velocities nearly that of light. The radiation traverses the 'Crab' Nebula virtually unimpeded, but the electrons are trapped by the strong magnetic fields in the nebula, and spiral their helter-skelter way around the field lines, losing energy in the form of radiation at optical, infrared and radio wavelengths. This synchrotron radiation is the soft yellowish light we see when we look at the Crab Nebula, and it makes the object one of the ten brightest radio sources in the sky.

The 'Crab' Nebula is a young object and it may be that all supernovae pass through a stage wherein the neutron star excites electrons to synchrotron radiation at optical wavelengths. Certainly it is the only one of its kind known and, moreover, there are others of not very different ages which do not behave in the same way. Whether it really is unique is therefore open to debate, but when you see that faint fuzzy blob floating in the field of view of your telescope, it is of some interest to know that it is the only synchrotron radiation nebula you will ever see.

10. EXTRAGALACTIC OBJECTS.

Our galaxy is not unique: other galaxies (spirals and irregulars, but not ellipticals) contain emission and reflection nebulae. From the latitudes of Europe the richest collection of extragalactic nebulae never rises, for these are the Magellanic Clouds, where every field of view contains several clusters, H II regions or planetary nebulae. Nonetheless there are galaxies near enough in the northern skies for their H II regions to be seen in even quite small telescopes. The brightest of these is NGC 604 in M33, and this is the object that should be tackled first. The offsets of all the H II in northern galaxies recorded in the NGC, measured from the centre of the parent galaxy are:

Messier 33	RA (mins arc)	Dec (mins arc)
NGC 588	3.7 p	0.6 S
592	2.1 p	0.8 S
595	0.9 p	2.1 N
604	2.5 f	7.5 N

The following IC nebulae are also M33 H II regions: 131, 132, 133, 134, 135, 136, 139, 140, 142 and 143.

Messier 101	RA (mins arc)	Dec (mins arc)
NGC 5447	1.7 p	4.5 S
5449	1.6 p	1.9 S *
5450	1.5 p	7.3 S *
5451	1.2 p	0.9 N *
5433	0.5 p	3.5 S *
5455	0.4 p	9.7 S
5458	0.0 f	5.4 S *
5461	1.2 f	1.8 S
5462	1.7 f	1.0 N
5471	3.1 f	2.6 N

Those marked with an asterisk are described as condensations in the galaxy, but are probably at least in part H II regions.

Each of these is particularly bright, much more so than any known in our galaxy; even the humble Large Magellanic Cloud tops our galaxy with the 30 Doradus complex (NGC 2070). Some more distant galaxies are known to contain even more luminous emission nebulae, and it has recently become difficult to draw a distinction between galaxies and H II regions, because some of the former are almost entirely composed of the latter. Galaxies clearly lie outside the scope of this volume of the Handbook, but it is perhaps worth mentioning that large numbers of emission-line galaxies are known. In some of these, such as M77 (NGC 1068), it is the nucleus which emits like an H II region, but other (fainter) examples - NGC 3690 and NGC 4385 for instance - show the entire galaxy to be like a giant H II region, and little if any of its light comes from solar-type stars.

11. SPECTROSCOPY.

Emission nebulae present the professional astronomer with a vast amount of data because their spectra contain a wealth of narrow emission lines. Many ratios of the intensities of these lines give information on the physical conditions - the temperature, density and chemical composition - of the nebula and of the illuminating star. So far only the hydrogen and helium lines have been described, but although these are by far the most abundant elements, their emission lines are overshadowed by those of heavier elements, notably singly-ionized oxygen lines at 3727 Å and the doubly-ionized oxygen pair at 4959 and 5007 Å. These lines are forbidden, which means that they cannot be observed in the high density environment of terrestrial laboratories, but thrive in the tenuous nebulae.

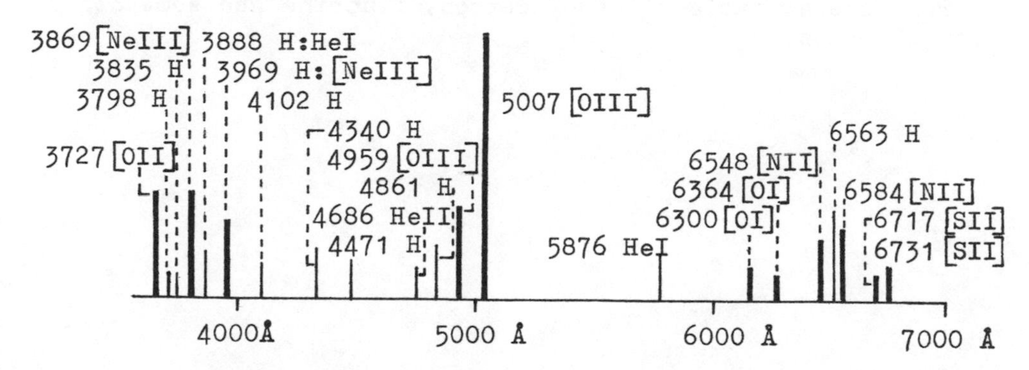

Figure 5. A Typical Emission Spectrum.

Why are the forbidden lines so strong? The simple answer is that less energy is involved in exciting them. The hydrogen and helium lines are part of a stack of linked emission lines which cumulatively require a sizeable amount of energy, typically ten times that emitted by each line, to excite them. On the other hand the forbidden lines are the very bottom of such a stack, so the energy required to excite them exactly equals the energy they radiate. So for a given input of ultraviolet energy equally shared, the forbidden lines can emit many more photons. The extra energy of the hydrogen and helium lines is emitted in the infrared and ultraviolet, where it cannot be seen. This, however, is only a partial answer, because it offers no explanation of why the lines which lie in the stack above the forbidden lines need not be excited, nor of why the energy is so equably shared.

Unlike the "permitted" lines of hydrogen and helium, the forbidden lines are not excited by direct ultraviolet radiation and subsequent cascading through the stack. Instead, the atoms and ions are raised directly to the appropriate energy levels by collisions with electrons, and from these levels they drop, emitting at their characteristic wavelengths. The electrons are produced by ionization of hydrogen and other atoms. If hydrogen is ionized by photons of shorter wavelength than is absolutely essential (911.5 Å), the extra energy of the photon is carried off by the electron and can be donated to one or more forbidden lines. Thus the electrons serve not only to excite just the relevant forbidden lines, but also to share the ultraviolet energy between the various emission lines.

Finally why do electrons not excite the forbidden lines of hydrogen and helium? Quite simply there are none. The lowest emission lines of their stacks are in the extreme ultraviolet, and these require more energy than a single electron can carry. Nor are there suitable lines of carbon, fluorine and some other common elements.

12. OBSERVING GASEOUS NEBULAE.

The conditions required for the successful observation of gaseous nebulae are similar to those needed for other types of nebulae - to wit, very dark skies, as large an aperture as possible and good dark adaption. However, there are a few other factors which deserve mention, and these depend upon the type of nebula under observation.

Reflection Nebulae.

These are usually very challenging because of the almost inevitable presence of at least one bright star in their midst. A reflection nebula is usually brightest near the star and fades in intensity with radial distance from it. In this respect it resembles the aura of light which surrounds any star (but it is most noticeable around bright stars) because of scattering in the atmosphere and telescope. You should therefore take special care to ensure that your optics are clean. Particularly beware of dewing. Nights of good seeing should be preferred, for if the bright central portion of the seeing disk is small, the amount of light at greater distances from the star will also be less. An occulting bar is a valuable piece of equipment if only one bright star is involved. A high magnification is usually to be preferred.

In some cases the nearby star is of considerable brightness. A good example is gamma Cas, near which are the reflection nebulae IC 59 and IC 63. Due to the proximity of this star, and the faintness of the two nebulae, observation will be rendered extremely difficult.

Always examine several nearby stars of similar magnitude to make sure that these too do not appear to be embedded in nebulae: legion are the spurious reflection nebulae which have been claimed.

Emission Nebulae.

Emission nebulae are usually brighter than reflection nebulae, are often much larger and can lie at considerable distances from their exciting stars. Thus the recommendations outlined above are not applicable. Emission nebulae are frequently larger than the field of view of large telescopes, some being larger than the moon. A low magnification is usually best to bring in some surrounding sky. Exceptions are small objects, an example being NGC 1931, which is detectable in a good sky at low power. In cases like this higher magnification can be utilised to try to define any structure the nebula may show.

A short-focal-length rich-field telescope is the ideal instrument; preferably a reflector for its colour rendition. The colour of nebulae depends upon their emission line intensities, and these may vary across an individual nebula. In M42, for instance, the centre appears green, but more distant features are red. Because most emission nebulae are green, a red light used at the telescope does not harm the relevant dark adaption, and can even be beneficial to the tired eye.

A direct vision spectroscope or prism is not useful for the detection of large emission nebulae because the emission lines overlap, due to the fact that each emission line forms an image of the entire nebula. Thus the nebula does not stand out against the stellar background any better than it would by direct vision. Only very small emission nebulae (usually planetary nebulae) can be found this way, as described in Part One. However, it is worth examining any nebula through a prism or spectroscope since the size and shape may vary from line to line. This too is best done at low magnification to minimise overlap of the different images. Note particularly the forbidden [O III] lines at 4959 and 5007 Å by comparison with the red H-alpha and forbidden [N II] group; the latter often come from a larger area, as is the case with the Orion Nebula where it can be seen directly by the increased redness of the outer portions.

While objects like M42 are both large and bright, many emission nebulae, while large, are of low surface brightness, and require the very best conditions. Particularly this applies to objects such as M8, the 'Lagoon' nebula in Sgr, which does not reach a great altitude in the light summer skies of the British Isles. For more southerly observers, however, this object presents a fine sight. Even so, in the British Isles a fair amount of detail can be seen in M8, of which mention will be made overleaf. A fainter object than M8 is M20, the 'Trifid' nebula, and although it is unlikely that the Trifid aspect will be seen from northern Europe, quite sizeable areas of nebulosity can be observed in good transparency.

Although M42, M8 and M17, the 'Omega' nebula, show different degrees of structural detail, the average amateur telescope will not resolve much, if any, detail in other nebulae, these appearing as no more than a glow fading off at the edges, this applying to even fairly bright nebulae like NGC 1491 or NGC 7129.

While objects in the IC are generally less easily observable visually than many NGC nebulae, some IC objects are, in fact, very easy to find. Nebulae like IC 432 and IC 435 in Orion are small but quite bright, and present no problems. Other IC objects in Orion, however, are difficult, examples being IC 424 and IC 426.

Orion also harbours other large nebulae, apart from M42, and objects like NGC 2024 will show structural details of quite considerable extent.

Dark Nebulae.

These are only as easy to see as are the bright backgrounds on which they are superimposed. In some cases, notably the dust clouds of the Milky Way, the backgrounds are produced by stars rather than glowing gas, so that a very low magnification (e.g., unity) is best. Any abrupt edge to a bright nebula must be indicative of one of two things: either the ionized gas is contained in a shell or bubble, as in a planetary nebula, or there is intervening dust. Particularly when observing H II regions, a dark patch or sharply defined edge is probably a dark nebula.

The dark nebula associated with M42 has already been mentioned, but there are two other nebulae, M8 and M17, where the association of bright and dark nebulae is to be seen. In M8 an area of dark nebulosity can be seen running off to the NW from the main nebula, and on a good night this is clear enough. The dark nebula that is connected with M17, however, is much greater in intensity, and can be seen on the W edge of the nebula, forming part of the 'Swan's Neck' aspect. Generally this dark wedge will be found to be darker than any part of the sky around this striking object.

Supernova Remnants.

The brightest supernova remnant is the 'Crab' Nebula. It will however, not reveal any detail to most telescopes in the amateur range. Larger apertures are needed to resolve structure, and with an aperture of 16½-inches brightness differences, as well as the 'bay' on the E edge of the nebula are visible.

The assorted NGC numbers that make up the Cygnus 'Loop' are of varying degrees of brightness, but traces of filamentary structure can be seen in telescopes of about 8-inches and up, while with larger instruments the filaments present a fine sight. The difference in width between the NGC 6960 and the NGC 6995 parts of the nebulosity are easily visible in most amateur telescopes.

The supernova remnant IC 443 in Gem is a much lower surface brightness object than the Cygnus 'Loop'. However, the brightest parts of the nebulosity are well away from the nearest bright star (eta Gem) and it is in this section of the nebulosity that traces of gas may be seen with a large reflector on nights of excellent transparency.

PART THREE : A CATALOGUE OF PLANETARY NEBULAE.

INTRODUCTION.

The catalogue contains observations of 80 planetary nebulae made by twelve observers using telescopes of 6 to 100-inches aperture. Distribution of data within the catalogue is as follows.

The extreme left sections of the pages show the Webb Society catalogue number (WS) followed by the actual designation of the nebula, (NGC, IC etc.). In the three cases where nebulae are Messier objects, the Messier number will be found below the relevant NGC number.

The remaining data, covering the greater parts of the respective pages, is detailed below.

Upper Line.

(a) Positions for 1975.0.

(b) Magnitudes of nebulae (m^n) followed by magnitudes of the central stars (m^s). A 'v' following a figure denotes a photovisual magnitude, the remainder being photographic. Where no magnitudes are available, surface brightness measures in red light are shown (SBr).

(c) Angular diameters of nebulae in arcseconds (AD). For some double-envelope nebulae the envelope of greater diameter will show on this line, the smaller diameter of the inner envelope being placed below, on the level of line two.

Second Line.

(a) Nebular types: classification is according to Vorontsov-Velyaminov (1934).

Type.	Description.
I	Stellar.
IIa	Oval, evenly bright, concentrated.
IIb	Oval, evenly bright, without concentration.
IIIa	Oval, unevenly bright.
IIIb	Oval, unevenly bright with brighter edges.
IV	Annular.
V	Irregular.
VI	Anomalous.

(b) Distances of nebulae in kpc (r), (Cahn and Kaler, 1971).

Third Line.

(a) Types of nuclei, (Gurzadyan, 1969; Perek and Kohoutek, 1969).
(b) Radial velocities (RV), (Perek and Kohoutek, 1969).
(c) The abbreviated form of the relevant constellation.

A Catalogue of Planetary Nebulae.

Visual Observations.

The data below the dotted line is concerned with the visual observations. The figures in parenthesis (60) (16½) etc., refer to the aperture used in inches, and this is followed by the observation. All observations are set out in order of decreasing aperture. Quoted magnitudes, diameters and distances in this section are purely visual estimates. Where observations made by more than one person using identical telescopes are concerned, the results have been amalgamated. The observations are all contracted renderings of initially more extended reports.

Field Drawings.

These will be found on the opposite pages to the written observations, and thus number four to a page. All are shown in circles of 2½-inches diameter, regardless of the actual field diameters, which are shown below each drawing along with details of telescopes used and the observers concerned. Orientation is north down, east to the right. For NGC 2452 and NGC 2610 no drawings were available, and field charts have been made up from photographs. For IC 1295 a field chart has been made up from a visual observation of the field, and is centred approximately on the position of this object.

A supplementary set of observations and drawings, made with a very wide variety of reflectors, can be found in Appendix 1.

List of Observers.

The following list shows the names of the observers whose work appears in the catalogue, plus details of their locations and respective telescopes.

D.A. Allen.	100-inch	Mt. Wilson, U.S.A.
	60	" " "
	60	Mt. Lemmon, "
	60	Teneriffe, Canary Islands.
	40	Sutherland, S.Africa.
	36	Herstmonceux, U.K.
	12	Coonabarabran, Australia.
J.K. Irving.	18	Salford Obs. U.K.
P. Arnott.	18	" "
M.J. Thomson.	16½	Santa Barbara, U.S.A.
G. Hurst.	10	Earls Barton, U.K.
J. Perkins.	10	Kirkby-in-Ashfield, U.K.
E.S. Barker.	8½	Herne Bay, U.K.
S.J. Hynes.	8½	Wistaston, "
C. Nugent.	8½	Upton, U.K.
K. Glyn Jones.	8	Winkfield, U.K.
P. Brennan.	8	Regina, Canada.
S. Selleck.	8	Santa Barbara, U.S.A.
K. Sturdy.	6	Helmsley, U.K.

A Catalogue of Planetary Nebulae.

Observers and Accredited Nebulae.

The following list shows all planetary nebulae that are listed in the catalogue, plus the initials of the respective observers alongside. The nebulae are listed in their catalogue order, i.e., in order of RA.

NGC	40	DAA, MJT, ESB.
	246	MJT.
	650/1	JKI, KGJ, KS.
IC	1747	DAA, ESB.
NGC	289	MJT, PB.
IC	351	DAA, ESB.
	2003	DAA, ESB.
NGC	1501	DAA, MJT, KGJ.
	1514	MJT, KGJ.
M	2-2	DAA, ESB.
NGC	1535	MJT, ESB.
J	320	DAA, ESB.
IC	418	MJT, ESB.
NGC	2022	GH, KGJ, KS.
IC	2149	JP, ESB.
	2165	ESB.
J	900	JP, ESB.
M	1-7	DAA, ESB.
NGC	2346	DAA, MJT, ESB.
	2371/2	JKI, ESB.
	2392	MJT, JP, KGJ, KS.
M	1-17	ESB.
NGC	2438	MJT, DAA, SS, KS.
	2440	KGJ.
	2452	MJT.
	2474/5	ESB.
	2610	SS.
	3242	MJT, KGJ, KS.
	3587	MJT, KGJ, KS.
	4361	MJT, KS.
IC	3568	ESB.
Me	2-1	DAA, PB.
IC	4593	ESB.
NGC	6058	MJT, ESB.
	6210	MJT, KGJ, KS.
	6309	ESB.
	6445	MJT, PB.
	6543	DAA, MJT, ESB.
	6572	DAA, JP.
	6567	PB.
M	1-59	DAA, PB.
Hu	2-1	DAA, ESB.
NGC	6720	DAA, MJT, KGJ.
IC	1295	MJT, PB.
NGC	6741	ESB.
	6751	DAA, MJT, ESB.
	6772	MJT.
IC	4846	ESB.
NGC	6778	ESB.
	6781	MJT, ESB, CN.
	6790	ESB.
	6803	ESB.
	6804	MJT, ESB, CN, KS.
	6807	ESB.
Me	1-1	ESB.
M	1-74	DAA, ESB.
NGC	6818	KGJ, KS.
	6826	DAA, MJT, KGJ, KS.
	6833	DAA, ESB.
	6842	MJT.
	6853	MJT, GH, KGJ.
	6857	MJT.
	6879	PB.
	6884	MJT, ESB.
	6886	ESB.
	6891	MJT.
	6894	MJT, PB, KS.
IC	4997	DAA, ESB.
NGC	6905	DAA, KGJ.
	7008	JP, ESB, KS.
	7009	DAA, MJT, KGJ, KS.
	7026	DAA, ESB.
	7048	PB.
IC	5117	DAA, ESB.
Hu	1-2	DAA, ESB.
NGC	7139	MJT.
IC	5217	DAA, ESB.
NGC	7293	MJT, PB.
	7662	DAA, MJT, ESB.
Hb	12	DAA, ESB.

WS	Cat.	RA	Dec	m^n	m^s	AD
1	NGC 40	00 11.7	+72 23	10.7	11.6pv	60 x 38
		Type: IIIb		r = 1.17		
		Star: WC8		RV -20.5		Cep.

(60) 50" diam.; uniform neb with central star.
(16½) Dark areas between star and edges; S.p. and S.f. edges brightest; star centrally set.
(8½) Slightly varied brightness; spectrum of star beaded with emission lines x204.

WS	Cat.	RA	Dec	m^n	m^s	AD
2	NGC 246	00 45.9	-12 17	8.0	10.5	240 x 210
		Type: IIIa		r = 0.74		
		Star: O7+dG/dF		RV -43		Cet.

(16½) Annular; S and S.p. edges brightest; brighter area in N.f. section; four stars involved, brightest one in the N.p. part.

WS	Cat.	RA	Dec	m^n	m^s	AD
3	NGC 650	01 40.3	+51 26	9.4v	16.6	157 x 87
	NGC 651	Type: V		r = 0.60		
	M76	Star: C		RV 123.6		Per.

(18) Two bright components, p. the brighter and triangular in shape; faint luminous arc S of these; diffuse glow surrounds.
(8) Hazy, nearly elliptical patch with waisted shape; PA 035°, diam. about 1'.5 x 1'.0.
(6) Bar of faint light joining two sections with faint envelope surrounding; magnifies well.

WS	Cat.	RA	Dec	m^n	m^s	AD
4	IC 1747	01 55.6	+63 12	13.0	15.0	13
		Type: IIIb		r = 2.11		
		Star: WC		RV -62.7		Cas.

(100) AD about 15"; slightly elliptical; green.
(8½) Slightly extended in PA 135° - 315°; very slight hint of darker centre.

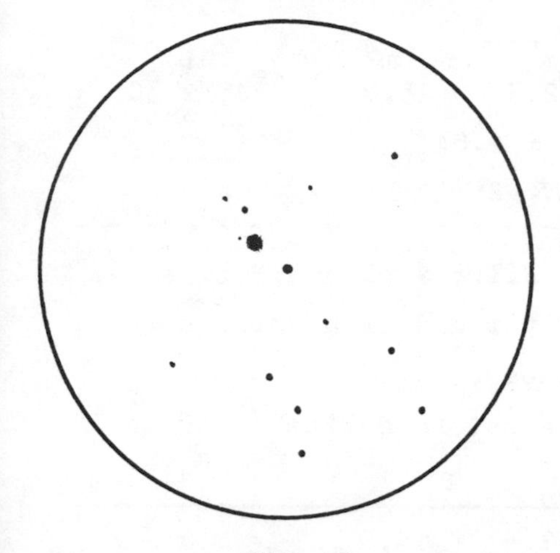
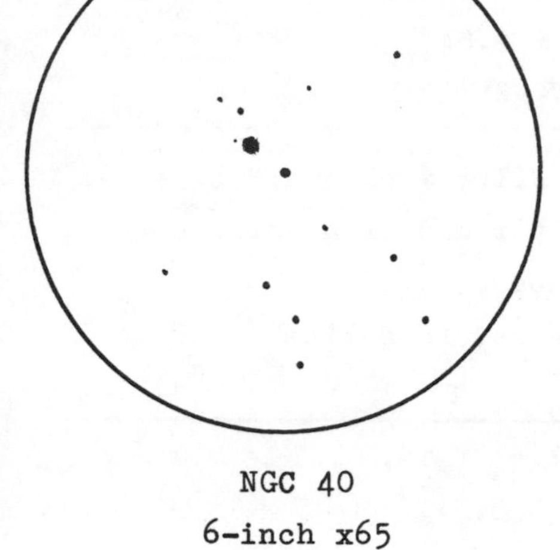

NGC 40
6-inch x65
Field 45'

P. Brennan

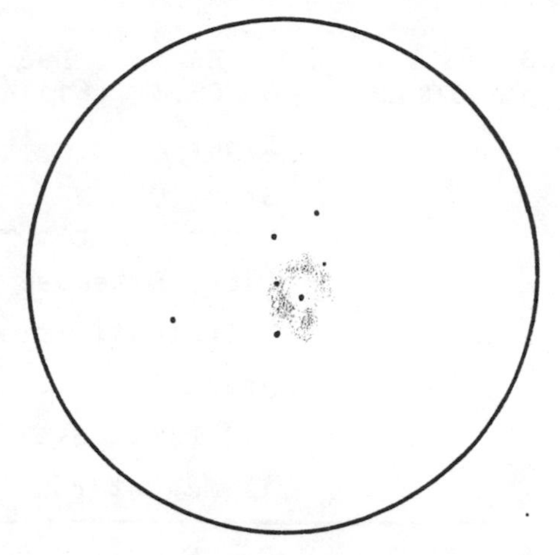

NGC 246
8-inch x75
Field 30'

P. Brennan

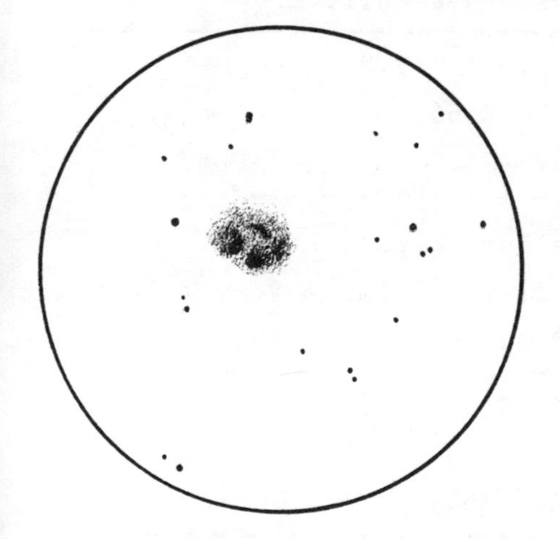

NGC 650-1
18-inch x200
Field 13'

J.K. Irving

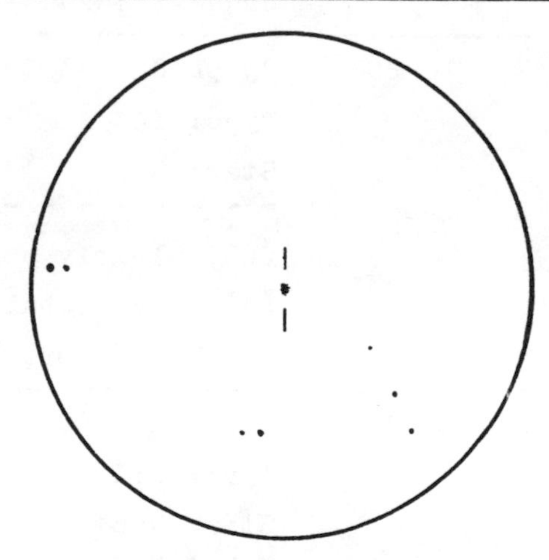

IC 1747
8½-inch x204
Field 12'

E.S. Barker

WS	Cat.	RA	Dec	m^n	m^s	AD
5	IC 289	03 08.3	+61 14	12.3	15.0	45 x 30

Type: IV   r = 1.84

Star: C   RV -28   Cas.

(16½) Extended N.p., S.f.; N.p. and S.p. edges brightest; irregularly round with various gradations of brightness.

(8) Faint oval nebula set in a line of 10 to 12 mag stars.

WS	Cat.	RA	Dec	m^n	m^s	AD
6	IC 351	03 45.9	+34 58	12.4	15.0	8 x 6

Type: IIa   r = 5.01

Star: C   RV -10.3   Per.

(100) Lozenge-shaped; AD 10 x 9"; no star.

(8½) PA 20° - 200°; definite oval x204 with even brightness and hazy edges; easy prism object; 8 and 9 mag stars 4' S.f.

WS	Cat.	RA	Dec	m^n	m^s	AD
7	IC 2003	03 54.8	+33 48	12.6	17.8	6

Type: II   r = 4.66

Star: C   RV -23.3   Per.

(60) Clearly non-stellar in poor seeing.

(8½) Hazy nebula x204; prism image very bright at this power; 9 and 11 mag pair 4' N.f.

WS	Cat.	RA	Dec	m^n	m^s	AD
8	NGC 1501	04 04.8	+60 51	10.0v	12.0	56 x 48

Type: III   r = 1.30

Star: WC6   RV +36.9   Cam.

(100) PA 100° - 280°; sharp outline showing fuzzy major axes: remainder uniform; star.

(16½) Annular; star visible over x100; x300 mottled with dark patches surrounding star.

(8) Hazy oval, slightly brighter centre; no star.

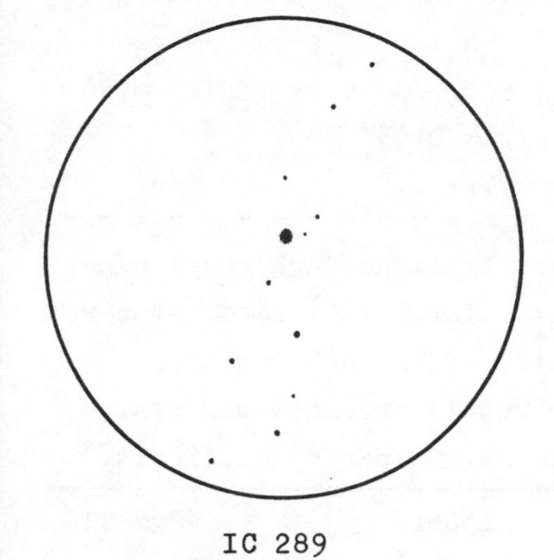

IC 289
8-inch x125
Field 25'

P. Brennan

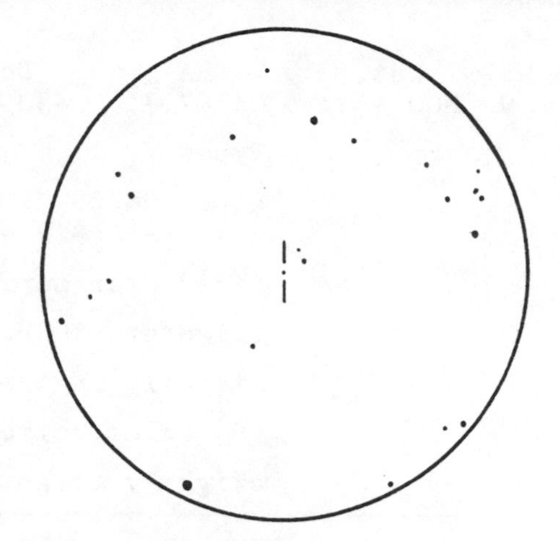

IC 351
$8\frac{1}{2}$-inch x51
Field 45'

E.S. Barker

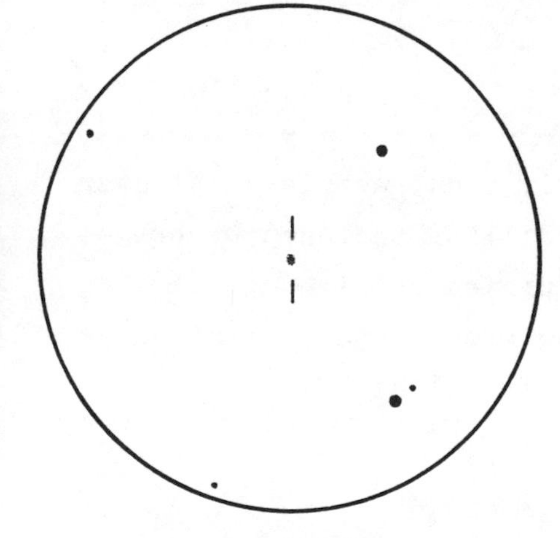

IC 2003
$8\frac{1}{2}$-inch x204
Field 12'

E.S. Barker

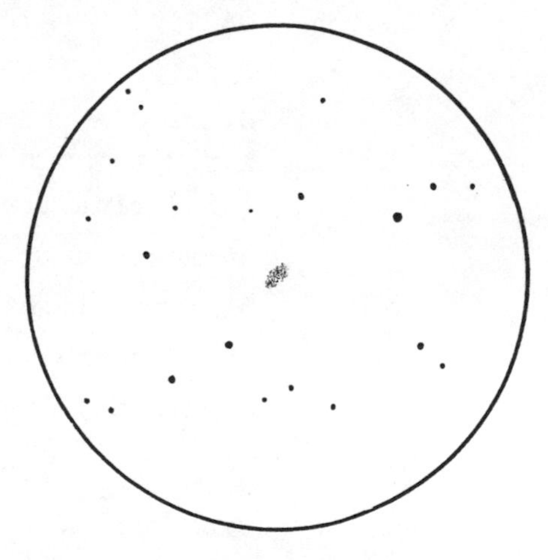

NGC 1501
8-inch x100
Field 40'

K. Glyn Jones

WS	Cat.	RA	Dec	m^n	m^s	AD
9	NGC 1514	04 07.7	+30 42	9.0	9.0v	120 x 90

Type: V　　r = 0.83
Star: 08/B9　　RV +36.9　　Tau.

(16½) Star surrounded by large dark ring; edges brighter at HP, particularly N.f. edge; star set directly in the centre of the nebulosity.
(8) Faint, circular nebula around 9 mag star; slightly brighter in NW, SE parts; difficult.

WS	Cat.	RA	Dec	m^n	m^s	AD
10	M2-2	04 11.2	+56 33	8.5SBr		12 x 11

Type: III　　RV -7　　Cam.

(60) Circular; soft edges and brighter centre.
(8½) Very faint blur x204; PA hard to define; just visible at MP in good transparency; not detectable with prism; very difficult.

WS	Cat.	RA	Dec	m^n	m^s	AD
11	NGC 1535	04 13.1	-12 48	9.6	10.0	20 x 17

Type: IV+VI　　r = 2.14
Star: C　　RV -1.4　　Eri.

(16½) Blue irregularly round nebula; x351 star surrounded by dark mottling enclosed by bright ring in slightly elongated shell.
(8½) Central mottling around star; faint outer nebulosity less symmetrical at HP.

WS	Cat.	RA	Dec	m^n	m^s	AD
12	J 320	05 04.1	+10 41	12.9	13.5	11 x 8

Type: II+IV　　r = 5.45　　8 x 5
Star: WN　　RV -23.4　　Ori.

(36) Green, non-stellar; diam. about 3".
(8½) x308 faint ellipse with ill-defined edges; star or slightly brighter centre faintly seen; fairly difficult, requiring use of prism at MP.

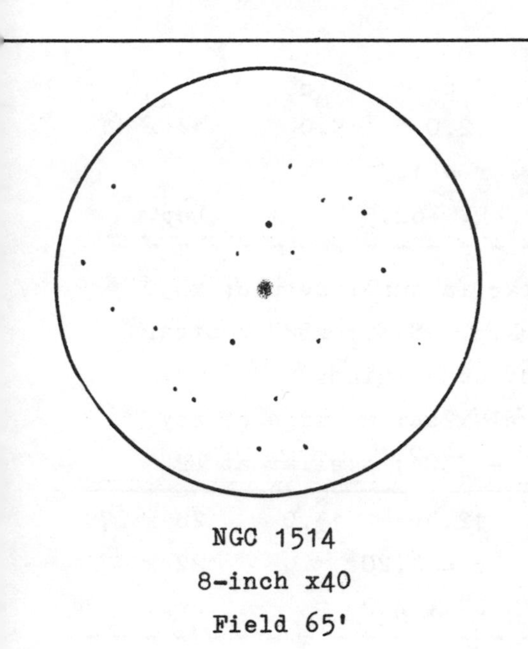

NGC 1514
8-inch x40
Field 65'

K. Glyn Jones

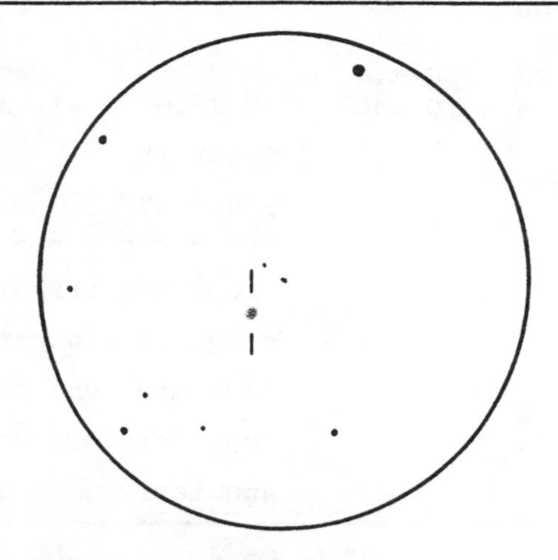

M2-2
$8\frac{1}{2}$-inch x204
Field 12'

E.S. Barker

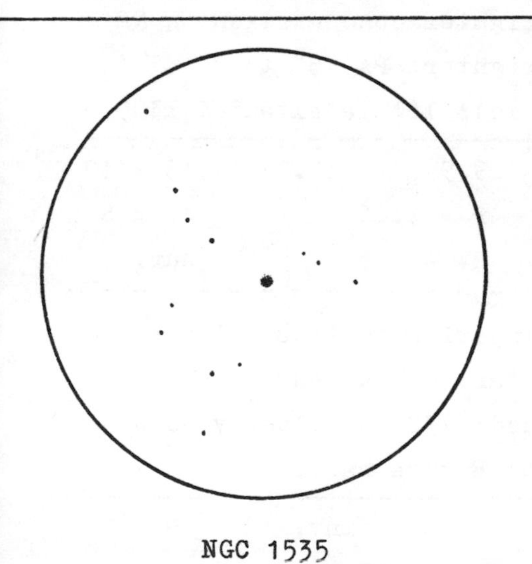

NGC 1535
8-inch x65
Field 41'

P. Brennan

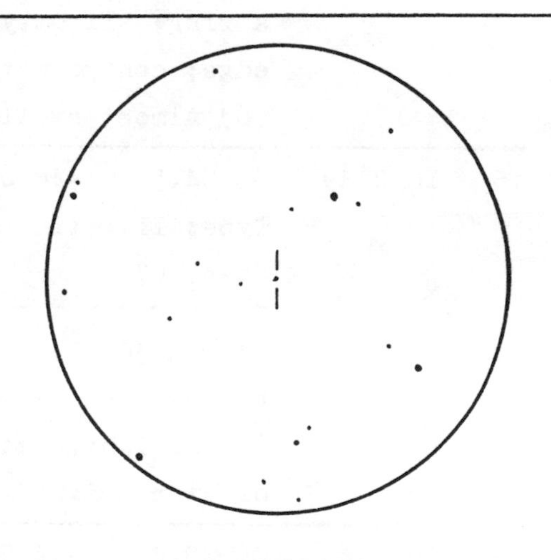

J 320
$8\frac{1}{2}$-inch x102
Field 24'

E.S. Barker

WS	Cat.	RA	Dec	m^n	m^s	AD
13	IC 418	05 26.6	-12 46	12.0	9.0	14 x 11
		Type: IV		r = 1.71		
		Star: 07		RV +62.5		Lep.

($16\frac{1}{2}$) x84 bright star in small nebula; x351 slightly elongated N.p., S.f.; x527 central dark area and bright outer ring.
($8\frac{1}{2}$) Very bright oval x204; no sign of any annularity; PA 150° - 330°; stellar at LP.

WS	Cat.	RA	Dec	m^n	m^s	AD
14	NGC 2022	05 40.7	+09 04	12.3v	13.0	28 x 27
		Type:IV+II		r = 2.20		22 x 17
		Star: C		RV +14.2		Ori.

(10) Fairly faint and stellar x40; x80 small round disk with no sign of central star.
(8) Oval patch showing slight impression of a ring; slightly brighter condensation on NE edge; centre not brighter; PA 15° to 20°.
(6) Almost stellar x61; little extended x305.

WS	Cat.	RA	Dec	m^n	m^s	AD
15	IC 2149	05 54.5	+46 07	9.9	10.2v	15 x 10
		Type: IIIb+II		r = 2.66		12 x 6
		Star: 07		RV -32.5		Aur.

(10) x130 bluish, tapering to fine points; bright centre and star; colourless at HP.
($8\frac{1}{2}$) Bright, extended; x308 nebulosity to W of star brighter, to E more hazy.

WS	Cat.	RA	Dec	m^n	m^s	AD
16	IC 2165	06 20.8	-12 58	12.5	inv.	9 x 7
		Type: IIIb		r = 3.87		
		Star: -		RV +55.3		CMa

($8\frac{1}{2}$) Bright prism image x102; x204 ellipse in PA 90° - 270°; easy object that can be identified without prism or spectroscope.

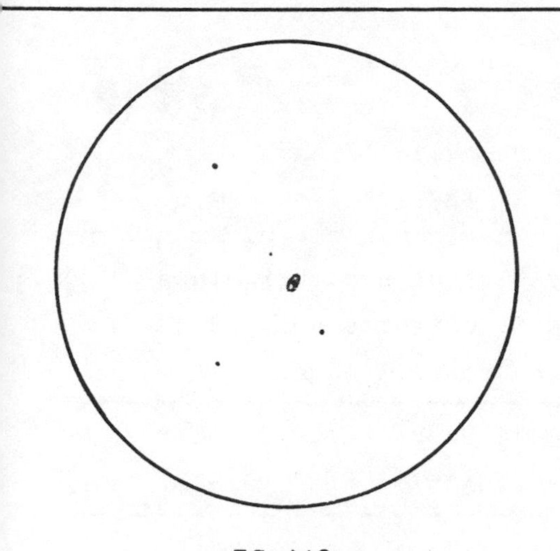

IC 418
18-inch x505
Field 6'

P. Arnott

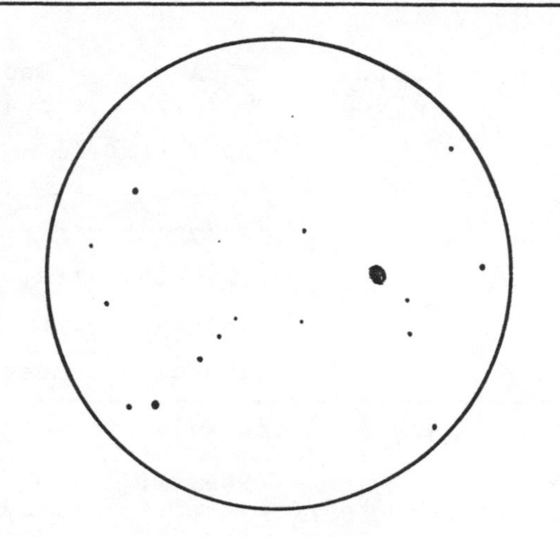

NGC 2022
8-inch x200
Field 20'

K. Glyn Jones

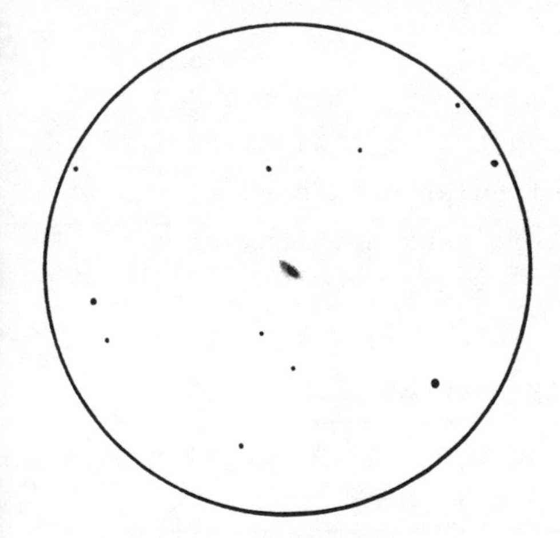

IC 2149
8½-inch x204
Field 12'

E.S. Barker

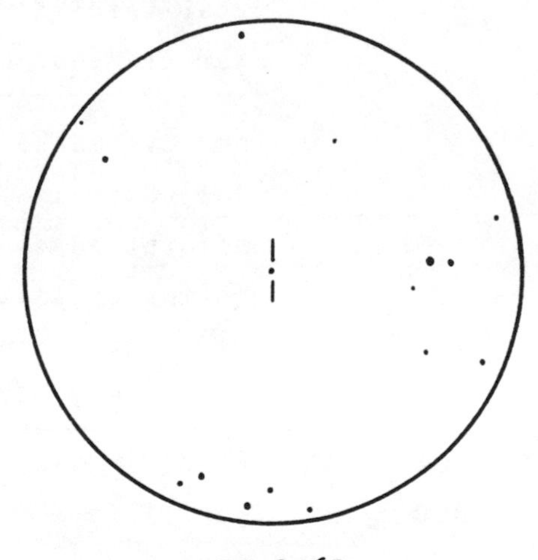

IC 2165
8½-inch x51
Field 45'

E.S. Barker

WS	Cat.	RA	Dec	m^n	m^s	AD
17	J 900	06 24.5	+17 48	12.4	inv.	12 x 10
		Type: IIIb+II		r = 2.16		
		Star: -		RV +47.2		Gem

(10) Elongated, grey blur of even brightness.
(8½) Easy prism object; definite oval x204 with suspected condensation near NE edge.

18	M1-7	06 37.2	+24 03	11.9SBr	inv.	38 x 20
		Type: II		RV +15		Gem.

(60) Circular; AD about 14"; no central star.
(8½) Elongated in PA 140° - 320°; overall even luminosity; requires good transparency and use of prism at MP; 9 mag star 40" NW, 13 mag star 40" E.

19	NGC 2346	07 08.1	-00 46	12.4SBr	10.6	60 x 50
		Type: IIIb+VI		r = ?		
		Star A+unseen comp.		RV +42		Mon.

(60) Oval nebula around bright blue star.
(16½) Star in faint circular nebula; N.f. and S.p. edges brighter; dark mottling also showing on latter.
(8½) Hazy featureless nebulosity; slight PA of 170° - 350°; easily seen at LP.

20	NGC 2371 NGC 2372	07 24.0	+29 32	13.0	12.5	54 x 35
		Type: IIIa+II		r = 1.70		
		Star: C		RV -20.3		Gem.

(18) Two almost circular sections, the N.f. being more diffuse and containing a double condensation.
(8½) Two nebulous patches in contact, SE the brighter; easy object even in slight haze.

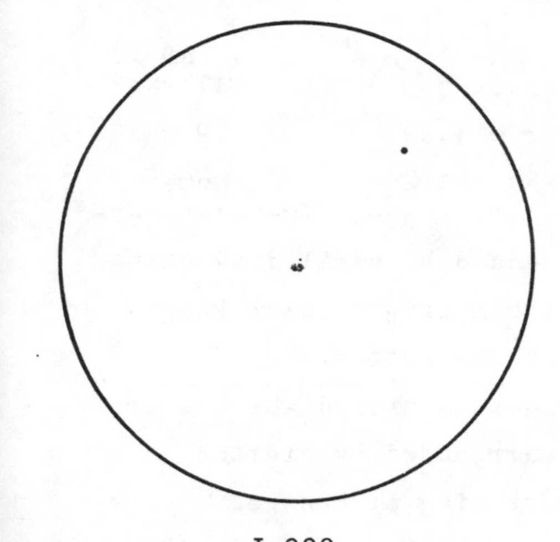

J 900
$8\frac{1}{2}$-inch x308
Field 6'

E.S. Barker

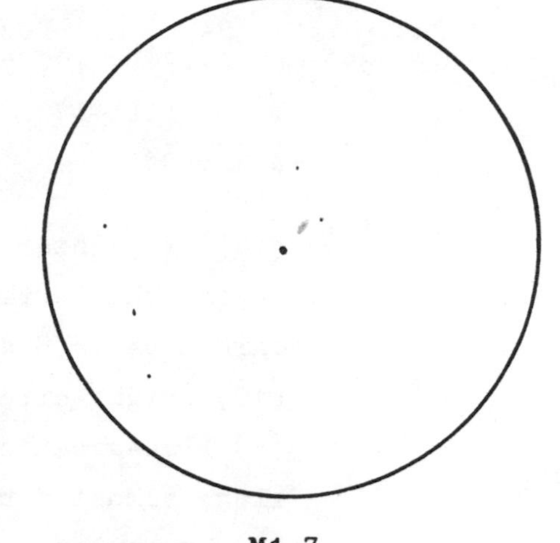

M1-7
$8\frac{1}{2}$-inch x204
Field 12'

E.S. Barker

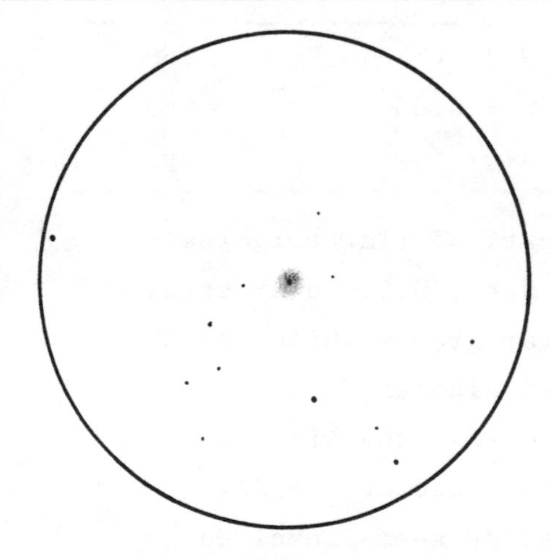

NGC 2346
$8\frac{1}{2}$-inch x204
Field 12'

E.S. Barker

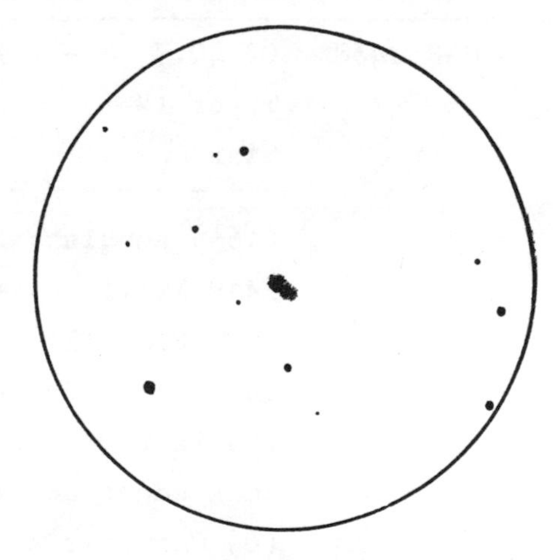

NGC 2371-2
18-inch x200
Field 13'

J.K. Irving

WS	Cat.	RA	Dec	m^n	m^s	AD
21	NGC 2392	07 27.7	+20 58	8.6v	9.1	47 x 43
		Type: IIIb+IV		r = 1.09		19 x 15
		Star: Of		RV +84.2		Gem.

(16½) x333 star surrounded by small disk and beyond this a ring within bright outer ring; dark area on S edge of the latter.
(10) Bright circular nebula around white star.
(8) Blue-green star surrounded by distinct ring; almost round dark ring at centre.
(6) Irregular brightness; star possibly a little to N; brightest part PA 160° - 290°.

WS	Cat.	RA	Dec	m^n	m^s	AD
22	M1-17	07 39.2	-11 30	-	-	3
		Type: I		RV +100		Pup.

(8½) Very faint prism image requiring HP; visible as a star x102; mag about 12.0.

WS	Cat.	RA	Dec	m^n	m^s	AD
23	NGC 2438	07 40.7	-14 40	9.7v	16.8	68
		Type: IV		r = 0.85		
		Star C		RV +77		Pup.

(16½) Annular; N.f. part of ring brightest; x419 little extended N.p., S.f.; star seen.
(12) Slightly irregular greeny-white patch on the NE edge of open cluster M46.
(8) Fairly large roundish patch with star on E edge; no trace of structure or PA.
(6) Circular at LP; at HP seems involved in star close S.f.

WS	Cat.	RA	Dec	m^n	m^s	AD
24	NGC 2440	07 41.0	-18 09	9.1v	inv.	54 x 20
		Type: V+III		r = 1.01		
		Star: -		RV +68		Pup.

(8) Bluish; almost circular centre in large halo; slightly elongated in PA 30° - 210°.

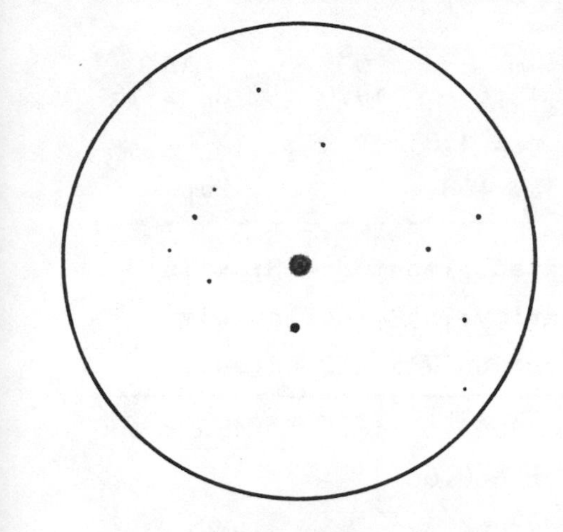

NGC 2392
8-inch x200
Field 20'

K. Glyn Jones

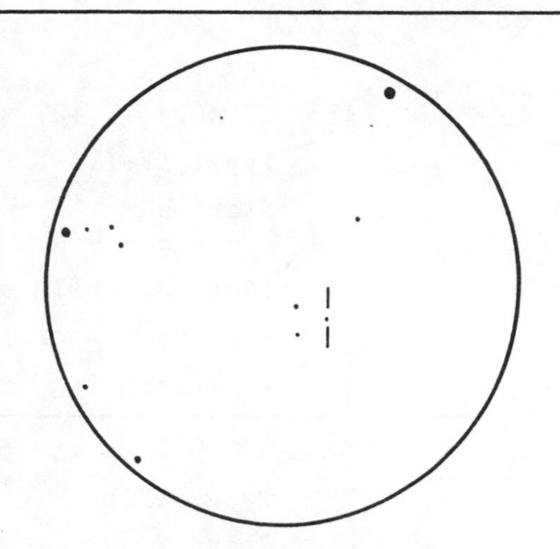

M1-17
8½-inch x102
Field 24'

E.S. Barker

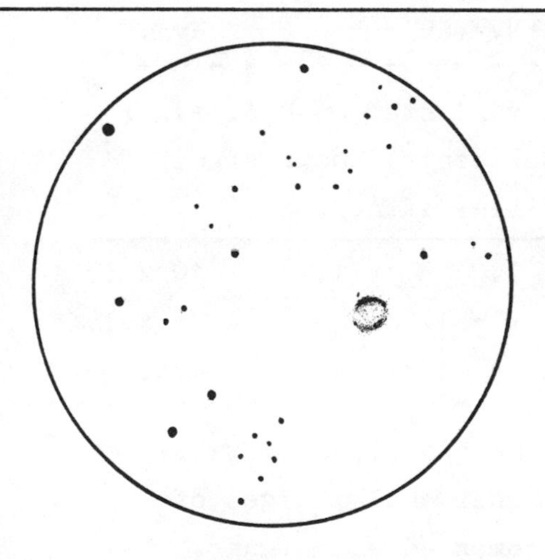

NGC 2438
18-inch x200
Field 13'

J.K. Irving

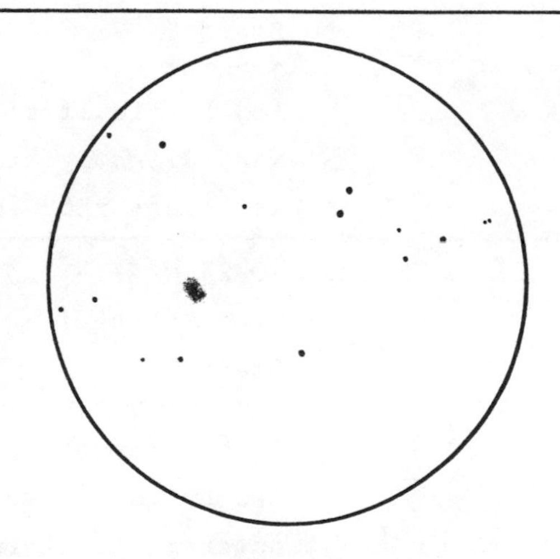

NGC 2440
8-inch x200
Field 20'

K. Glyn Jones

WS	Cat.	RA	Dec	m^n	m^s	AD
25	NGC 2452	07 46.4	+27 17	12.6	19.0	22 x 16
		Type: IV+III		r = 1.83		
		Star: WR		RV +68		Pup.

(16½) Slightly elongated almost N - S; x176 suggestions of annularity; x527 definitely annular; ring brightest on N and S edges.

WS	Cat.	RA	Dec	m^n	m^s	AD
26	NGC 2474	07 56.0	+52 56	14.0	16.6	450 x 400
	2475	Type: IIIb		r = 0.60		
		Star: ?		RV +2		Lyn.

(8½) One section visible x51 3' SW of 9 mag star; x204 clear image with oval shape and blurred edges; no sign of other component.

WS	Cat.	RA	Dec	m^n	m^s	AD
27	NGC 2610	08 32.4	-16 03	13.6	15.5	38 x 31
		Type: IV+II		r = 1.92		
		Star: ?		RV +88		Hya.

(8) Difficult at LP; x121 star on N.f. edge; x145 slightly brighter centre shows signs of mottling; x362 edges ill-defined.

WS	Cat.	RA	Dec	m^n	m^s	AD
28	NGC 3242	10 23.6	-18 31	9.0	11.7	40 x 35
		Type: IIIb+IV		r = 1.08		26 x 16
		Star: C		RV +4.7		Hya.

(16½) Very blue, extended N.p., S.f.; x527 two bright spots in inner shell; edges of outer shell diffuse compared with inner.
(8) Bright blue oval, brighter to NE and N; dark annulus around star; faint traces of outer ring; 10 mag star 5' S.f.
(6) Green-blue at LP; elliptical centre and round outer nebulosity; PA 150°.

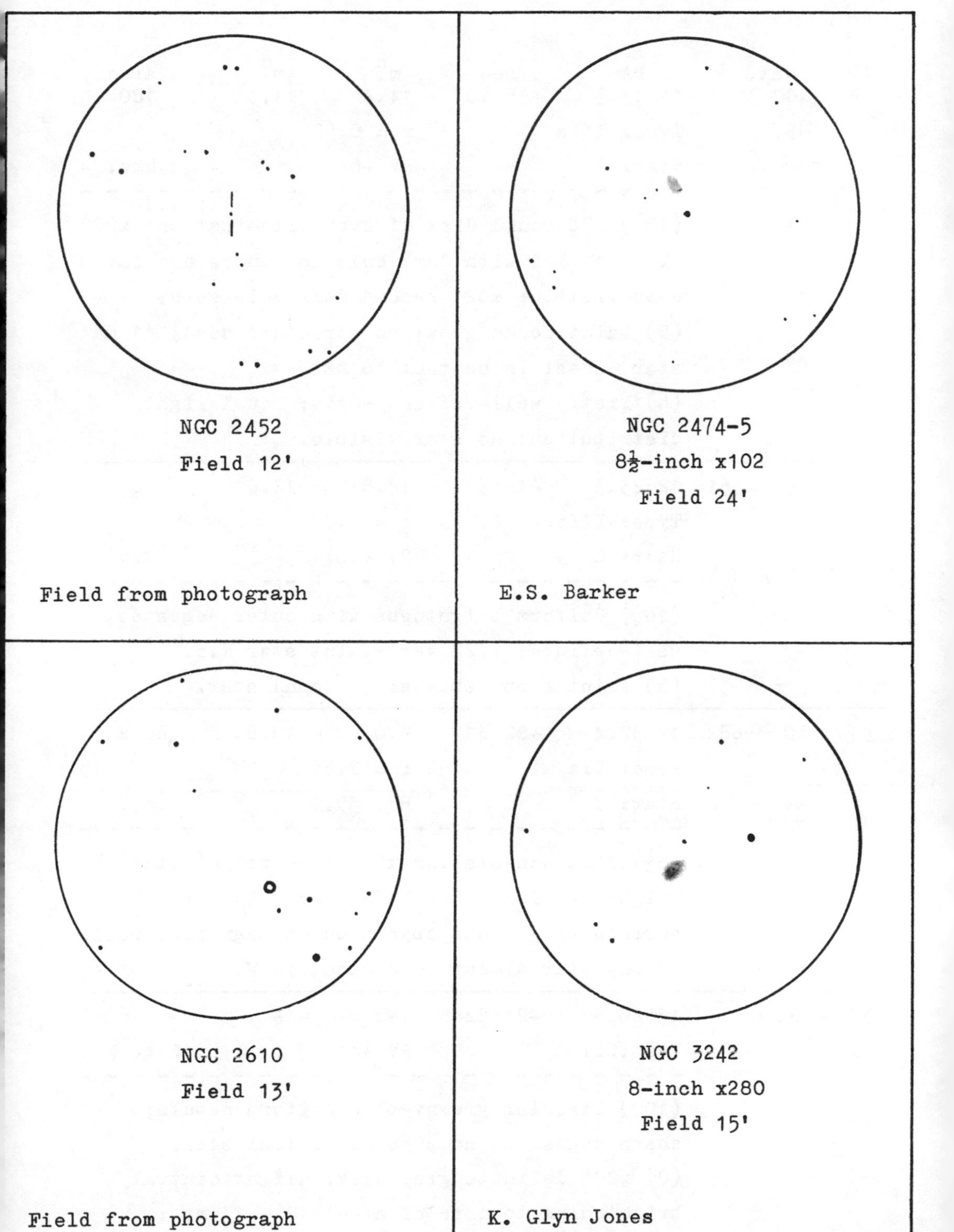
NGC 2452
Field 12'
Field from photograph
NGC 2474-5
8½-inch x102
Field 24'
E.S. Barker
NGC 2610
Field 13'
Field from photograph
NGC 3242
8-inch x280
Field 15'
K. Glyn Jones

WS	Cat.	RA	Dec	m^n	m^s	AD
29	NGC 3587	11 13.5	+55 10	11.4	14.3	180
	M97	Type: IIIa		r = 0.52		
		Star: ?		RV +8		UMa.

(16½) x70 round disk of even illumination; x222 disk mottled with dark hole in centre and faint star visible; x421 second dark hole seen.

(8) Faint round glow; no structure seen; 11 mag star almost in contact to NE.

(6) Pretty well-defined edges; equal light distribution; no star visible.

WS	Cat.	RA	Dec	m^n	m^s	AD
30	NGC 4361	12 23.3	-18 38	12.8	12.8	81
		Type: IIIa		r = 1.02		
		Star: C		RV +10		Crv.

(16½) Uniform brightness with outer edges not well-defined; x222 very faint star N.p.

(6) Faint even nebulosity around star.

WS	Cat.	RA	Dec	m^n	m^s	AD
31	IC 3568	12 32.4	+82 41	9.0	10.8	40 x 35
		Type: IIa		r = 2.65		15
		Star: C		RV -39.9		Cam.

(8½) Just non-stellar x51; x308 bright star in bright nebulosity with fainter halo; high surface brightness object which magnifies well; 13 mag star almost in contact to W.

WS	Cat.	RA	Dec	m^n	m^s	AD
32	Me2-1	15 20.9	-23 22	-	-	6
		Type: II		RV +44		Lib.

(100) Circular greeny-blue uniform nebula; sharp edges and no sign of central star.

(8) x205 definite grey disk; slight central brightening to core of about 12 - 13 mag; possibly slightly elongated N - S.

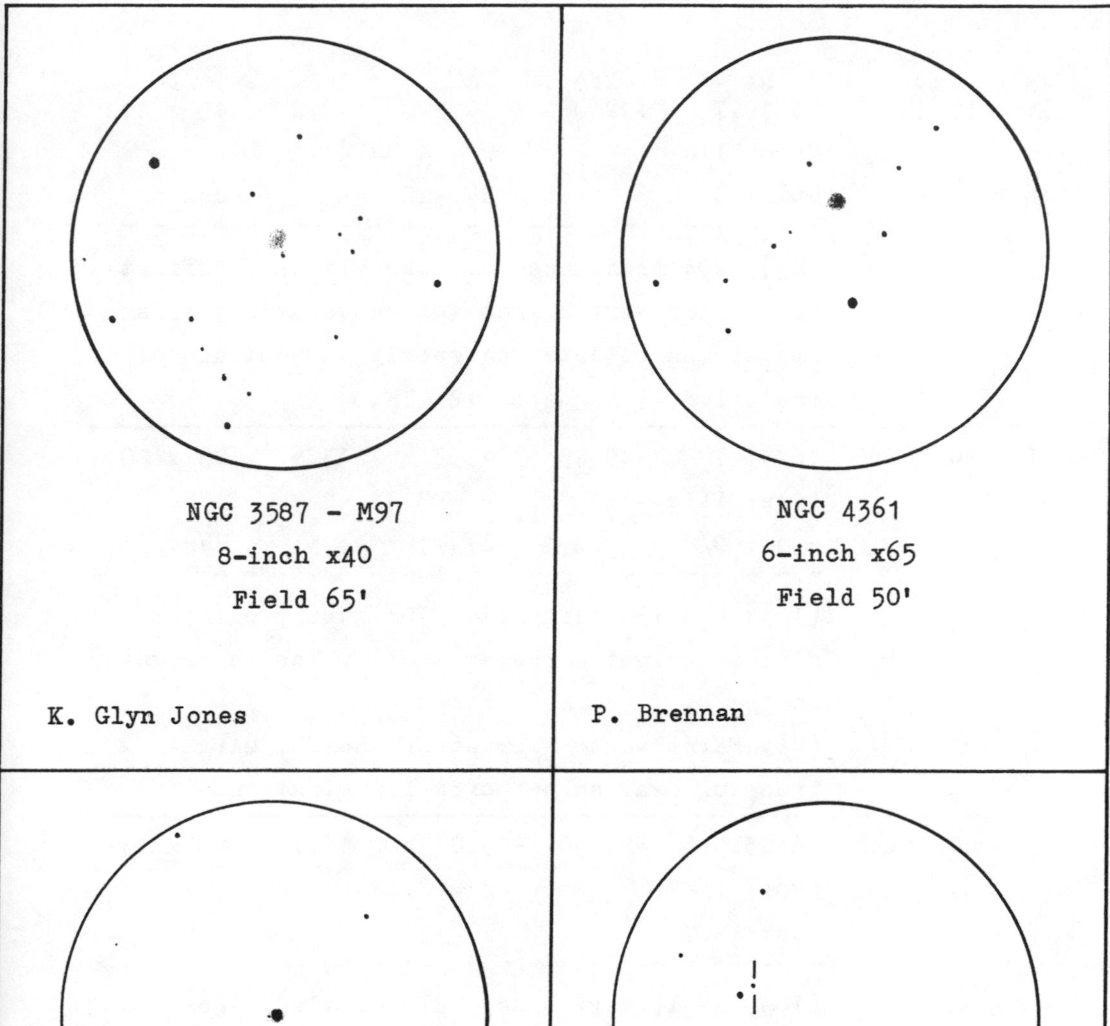

NGC 3587 - M97
8-inch x40
Field 65'

K. Glyn Jones

NGC 4361
6-inch x65
Field 50'

P. Brennan

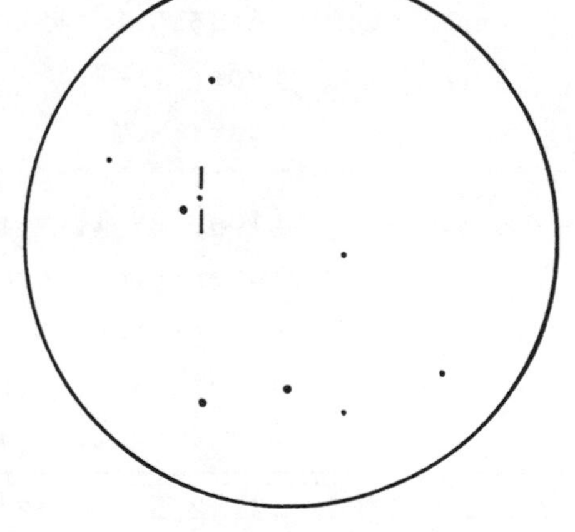

IC 3568
8½-inch x102
Field 24'

E.S. Barker

Me2-1
8-inch x75
Field 35'

P. Brennan

WS	Cat.	RA	Dec	m^n	m^s	AD
33	IC 4593	16 10.7	+12 08	10.2	10.2	15 x 11

Type: IIa r = 4.32

Star: Of RV +22 Her.

(8½) x204 pronounced ellipse with star offset to E; star very bright and shows strong prism image; can be detected easily without use of the prism; 9 mag star 12' SE.

WS	Cat.	RA	Dec	m^n	m^s	AD
34	NGC 6058	16 13.7	+40 45	13.3	13.5	25 x 20

Type: IIIa r = 3.41

Star: O/C RV +1 Her.

(16½) Easily recognized x70; x160 prominent star in bright circular shell which is brighter on the N.p. and S.f. edges.

(8½) Faint vague blur at MP and HP; slight trace of oval shape; even illumination.

WS	Cat.	RA	Dec	m^n	m^s	AD
35	NGC 6210	16 43.5	+23 50	9.7	11.7	20 x 13

Type: II+VI r = 2.04

Star: WC7 RV +1 Her.

(16½) Oval, very blue; x222 mottled edges and outer ring seen x400; takes HP well.

(8) Bluish; bright centre about 4" in halo about twice central diameter; PA 80° - 260°.

(6) Bright, small, blue; diam. about 30".

WS	Cat.	RA	Dec	m^n	m^s	AD
36	NGC 6309	17 12.6	-12 53	9.7v	13.0	19 x 10

Type: IIIb+VI r = 3.51

Star: C RV +91 Oph.

(8½) Small, faint due to low altitude; x204 elongated and slightly mottled image; faint star at each end of major axis; takes high magnification well.

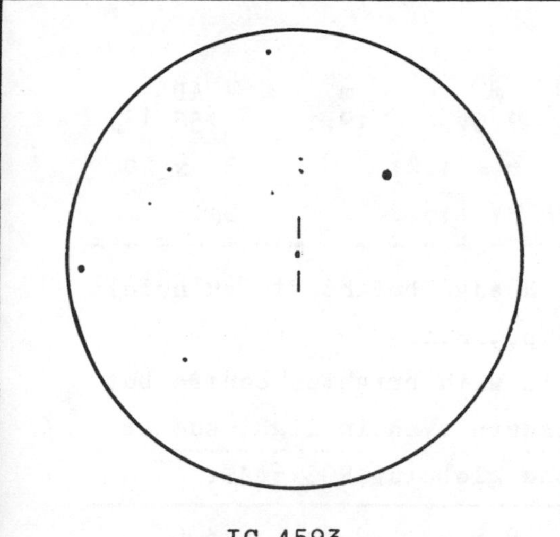

IC 4593
8½-inch x51
Field 45'

E.S. Barker

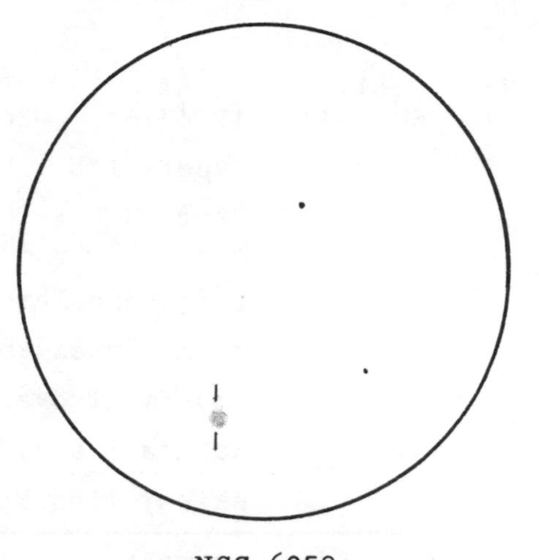

NGC 6058
8½-inch x204
Field 12'

E.S. Barker

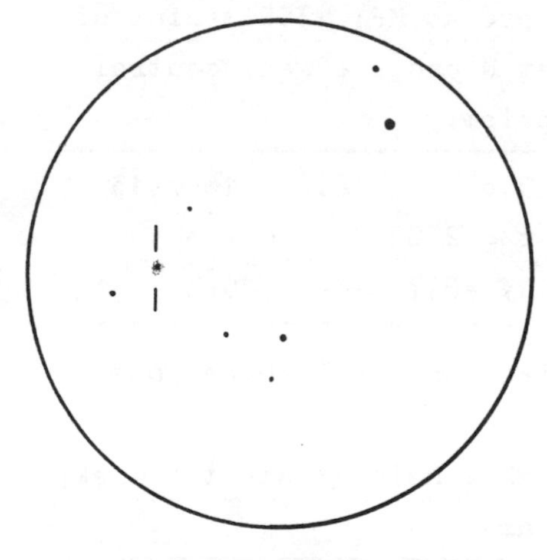

NGC 6210
8-inch x200
Field 20'

K. Glyn Jones

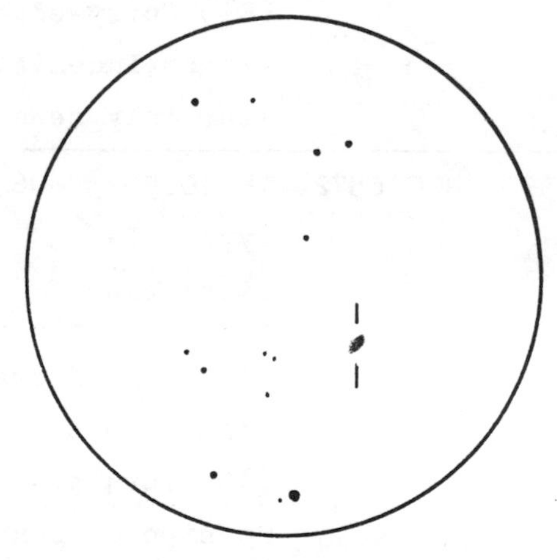

NGC 6309
6-inch x65
Field 50'

P. Brennan

WS	Cat.	RA	Dec	m^n	m^s	AD
37	NGC 6445	17 47.8	-20 00	9.5v	19.1	150
		Type: IIIb		r = 1.81		38 x 29
		Star: C		RV +16.2		Sgr.

(16½) Annular x160; N edge brightest and nebula slightly extended N.p., S.f.
(8) Small oval nebula with brighter centre but no star; easy to discern even in light summer skies; lies NE of the globular NGC 6440.

WS	Cat.	RA	Dec	m^n	m^s	AD
38	NGC 6543	17 58.6	+66 38	8.8	9.6v	300
		Type: IIIa		r = 1.52		22 x 16
		Star: WC+WN		RV -65.7		Dra.

(60) AD 40"; pale green uniform nebula; star.
(16½) Star in bright nebula within fainter shell; x160 star in dark area; inner ring.
(8½) Fuzzy-edged ellipse at MP; x308 a hint of slight indentations at N and S edges; central star only seen with prism.

WS	Cat.	RA	Dec	m^n	m^s	AD
39	NGC 6572	18 10.9	+06 50	9.6	12.0	16 x 13
		Type: IIa		r = 2.05		
		Star: WN6		RV -8.7		Oph.

(100) Vivid green, like two overlapping comet heads; central star.
(10) Turquoise disk; x232 halo around the disk; no sign of central star.

WS	Cat.	RA	Dec	m^n	m^s	AD
40	NGC 6567	18 12.3	-19 05	11.7	15.5	11 x 7
		Type: IIa		r = 2.21		
		Star: C		RV +119.8		Sgr.

(8) x125 slightly nebulous star; x310 slight elongation in PA 20° - 200°; 12 mag star almost in contact to E; in very rich field.

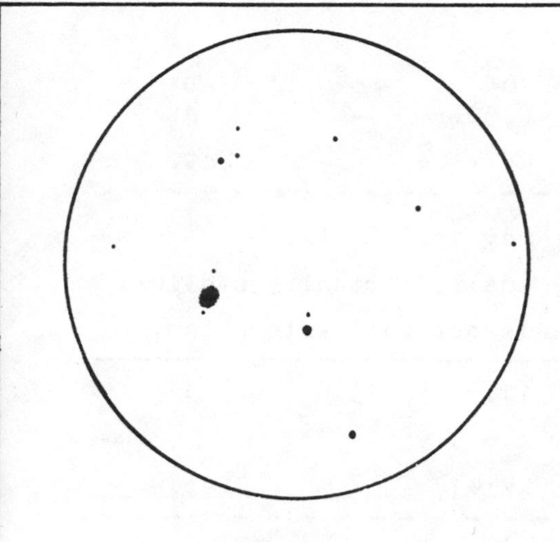

NGC 6445
8-inch x125
Field 21'

P. Brennan

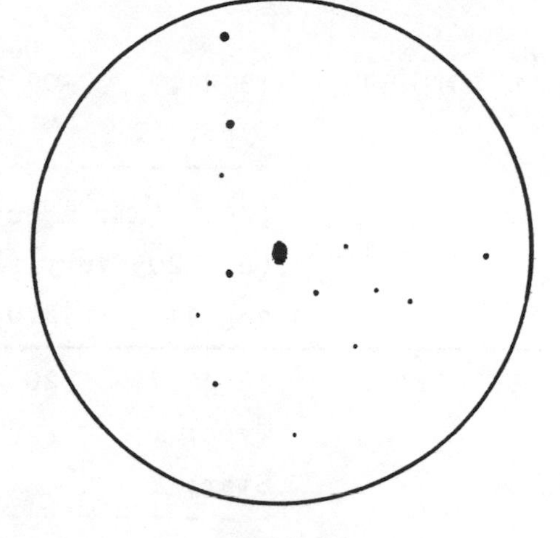

NGC 6543
6-inch x65
Field 41'

P. Brennan

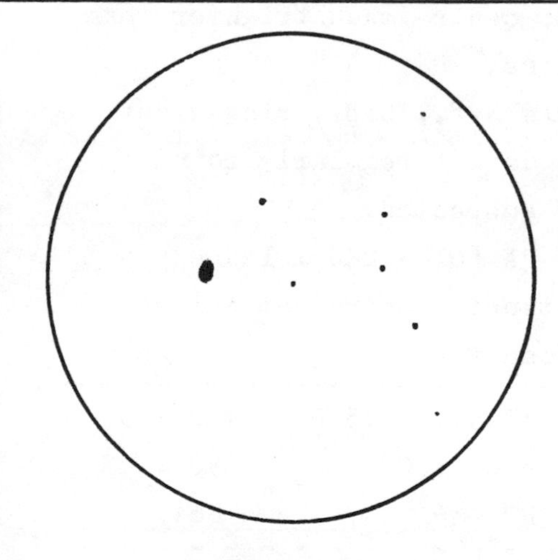

NGC 6572
8-inch x280
Field 15'

K. Glyn Jones

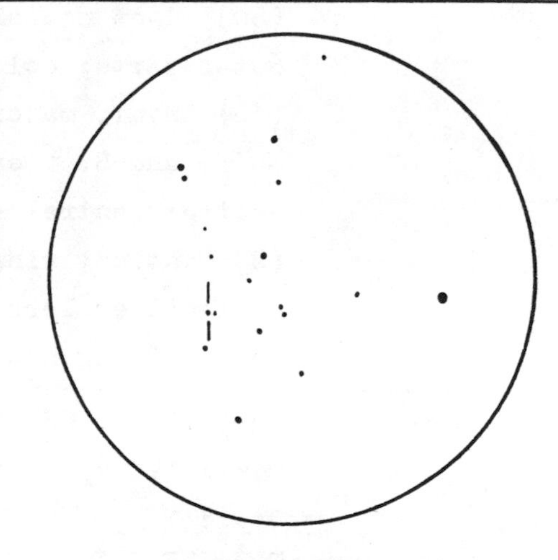

NGC 6567
8-inch x125
Field 25'

P. Brennan

WS	Cat.	RA	Dec	m^n	m^s	AD
41	M1-59	18 41.9	-09 05	6.5SBr	-	4
		Type: II		RV +104		Sct.

(40) Diam. about 3"; mag 13.
(8) x205 very faint, small; virtually stellar; mag 11.5 - 12.0; stands out well with prism.

WS	Cat.	RA	Dec	m^n	m^s	AD
42	Hu2-1	18 48.7	+20 49	11.5	13.0	3
		Type: I		r = 6.95		
		Star: ?		RV +17		Her.

(100) Green; diam. about 2".
(8½) Bright prism image; stellar on all powers.

WS	Cat.	RA	Dec	m^n	m^s	AD
43	NGC 6720	18 52.6	+33 00	8.9v	14.0	85 x 62
	M57	Type: IV		r = 0.78		
		Star: C		RV 122.3		Lyr.

(60) Edges not sharp; centre much brighter than outer parts; colourless.
(16½) Oval, major axis N.f., S.p.; ring wider on f. and S.p. ends; dark irregularly round centre; central star suspected.
(8) Distinct ring in PA 60° - 240°; longer extremities less distinct; centre not quite dark; 12 mag star close f.

WS	Cat.	RA	Dec	m^n	m^s	AD
44	IC 1295	18 53.2	-08 51	13.5v	15.0	120 x 90
		Type: IV		r = 1.20		90 x 83
		Star: ?		RV -36		Sct.

(16½) Very diffuse; two stars involved; x351 quite mottled; 25' S.f. globular NGC 6712.
(8) Nebulous spot with slight central brightening; diam. about 1'.0.

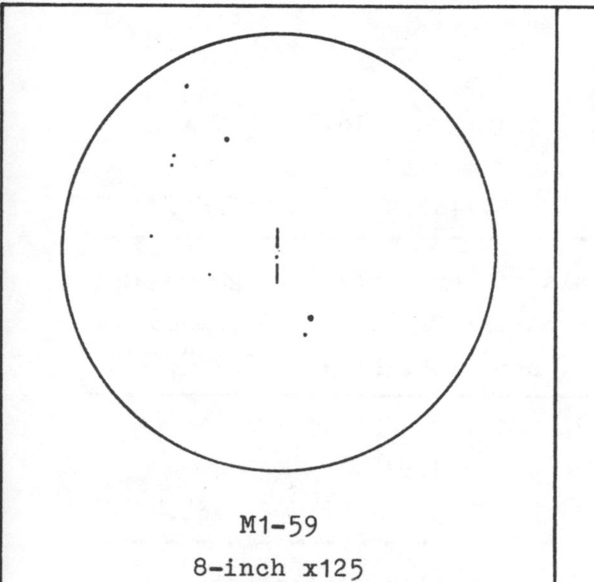

M1-59
8-inch x125
Field 20'

P. Brennan

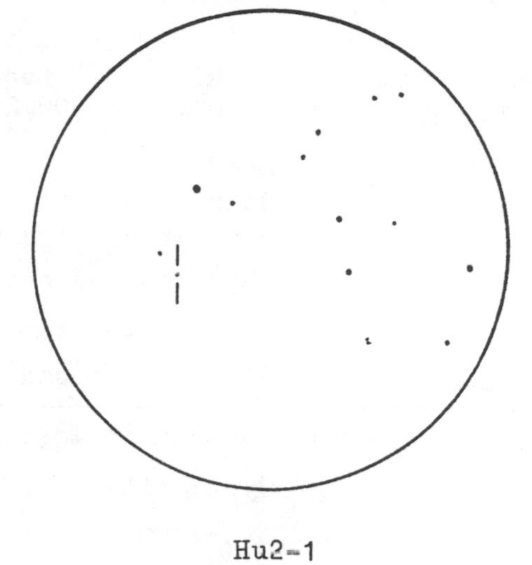

Hu2-1
8-inch x75
Field 35'

P. Brennan

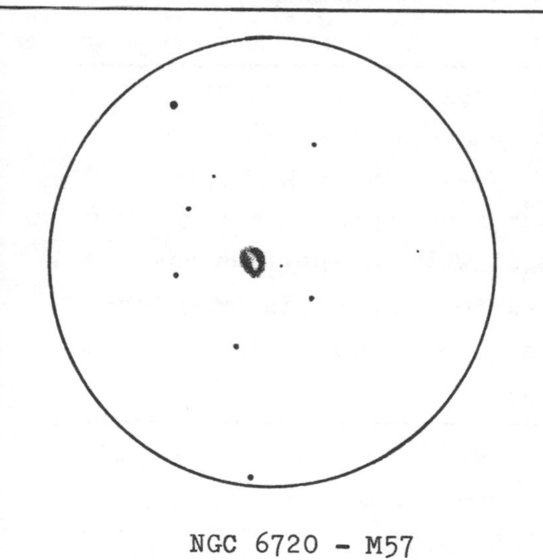

NGC 6720 - M57
8-inch x200
Field 20'

K. Glyn Jones

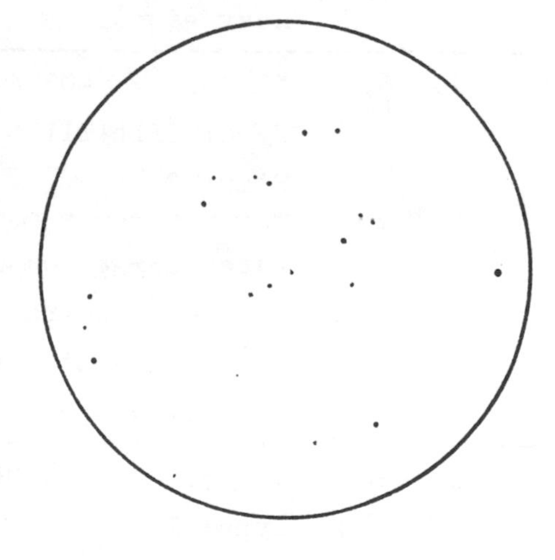

IC 1295
8½-inch x51
Field 45'

E.S. Barker

WS	Cat.	RA	Dec	m^n	m^s	AD
45	NGC 6741	19 01.3	-00 29	10.2v	16.7	9 x 7

Type: IV r = 2.57

Star: ? RV +42.9 Aql.

- -

($8\frac{1}{2}$) Blurred nebula of low surface brightness; faint prism image at LP; PA 90° - 270°; lies 5' S of two stars, sep. 30", PA 90° - 270°, 8^m.

46	NGC 6751	19 04.5	-06 02	11.0v	13.3	18

Type: III r = 1.91

Star: WC6 RV -36 Aql.

- -

(100) Diam. 18"; annular with central star.

($16\frac{1}{2}$) Fairly bright and round; best observed x222 and x333 which shows brighter centre but no central star.

($8\frac{1}{2}$) Even illumination x102; clearly round with hazy edges.

47	NGC 6772	19 13.3	-02 44	14.2	18.1	75 x 56

Type: IIIb+III r = 1.35

Star: ? RV ? Aql.

- -

($16\frac{1}{2}$) Large round spot with ill-defined edges; HP shows central area to be a little brighter; no central star seen; appearance similar to that of NGC 3587 (M97).

48	IC 4846	19 15.0	-09 06	12.7	16.3	2

Type: I r = 10.38

Star: ? RV +151 Aql.

- -

($8\frac{1}{2}$) Faint nebula seen with prism only at powers over x200; completely stellar when viewed normally; 11 mag and 13 mag stars, sep. 1'.0, about 7'.0 to E.

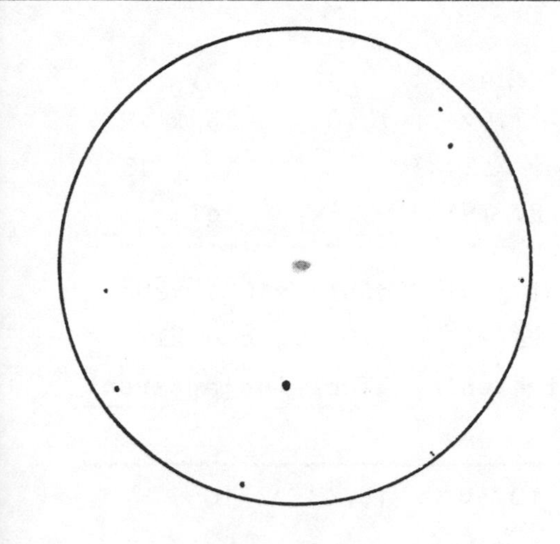

NGC 6741
8½-inch x204
Field 12'

E.S. Barker

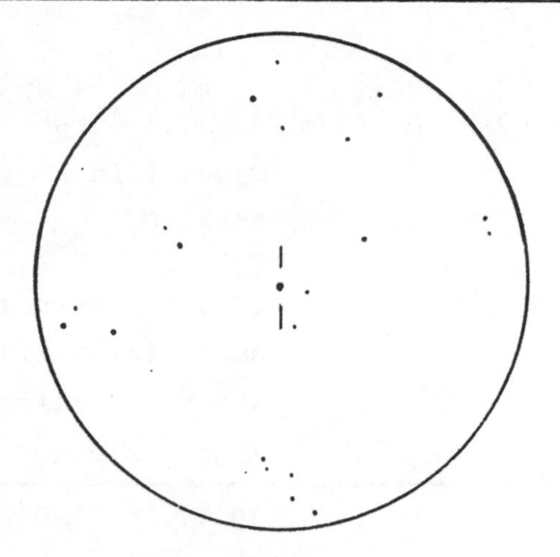

NGC 6751
8½-inch x102
Field 24'

E.S. Barker

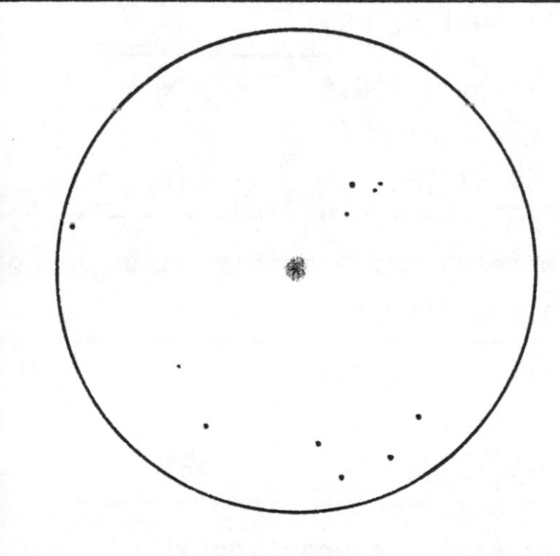

NGC 6772
8½-inch x102
Field 24'

E.S. Barker

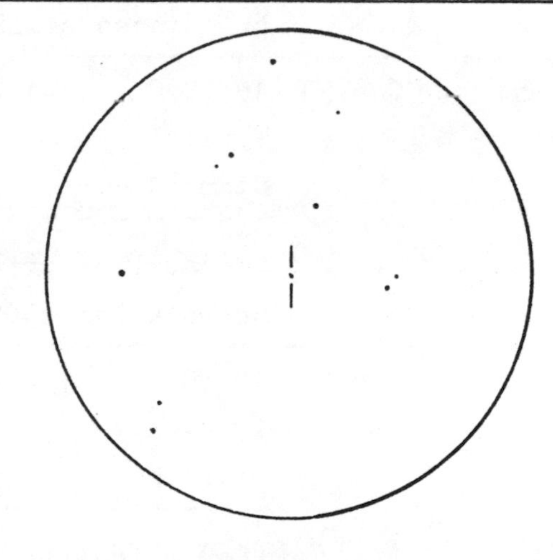

IC 4846
8½-inch x102
Field 24'

E.S. Barker

WS	Cat.	RA	Dec	m^n	m^s	AD
49	NGC 6778	19 17.1	-01 37	12.6v	15.0	25 x 19
		Type: IIIa		r = 2.43		19 x 13
		Star: C		RV +91		Aql.

($8\frac{1}{2}$) Not seen x51 but just visible x102; x204 small faint blur in PA 90° - 270° with oval shape and slight brightening along the apparent major axis.

50	NGC 6781	19 17.2	+06 30	10.3v	15.4	106
		Type: IIIa		r = 0.91		
		Star: ?		RV +6		Aql.

($16\frac{1}{2}$) Annular at x160; S edge brightest; no central star; star close to N.f. edge.
($8\frac{1}{2}$) Large, fairly diffuse; suspected signs of annularity; E edge a little more defined; surface brightness similar to that of M97.

51	NGC 6790	19 22.0	+01 28	10.3v	10.5	9 x 5
		Type: I		r = 6.21		
		Star: C		RV +41.8		Aql.

($8\frac{1}{2}$) Bright prism image at HP; possibly just non-stellar x308; 11 mag star 30" N.p.

52	NGC 6803	19 30.1	+10 00	11.4	14.1	5
		Type: IIa		r = 3.11		
		Star: WC		RV +13.1		Aql.

($8\frac{1}{2}$) Bright prism image x51; comparison with field stars x308 showed nebula as a definite disk with possibly slightly fuzzy edges; set in rich field.

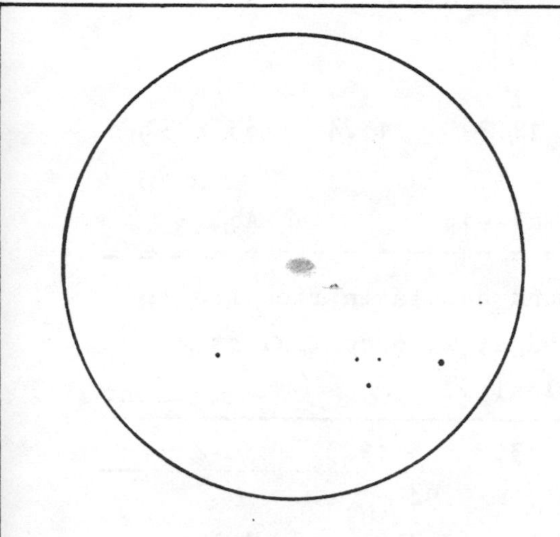

NGC 6778
$8\frac{1}{2}$-inch x204
Field 12'

E.S. Barker

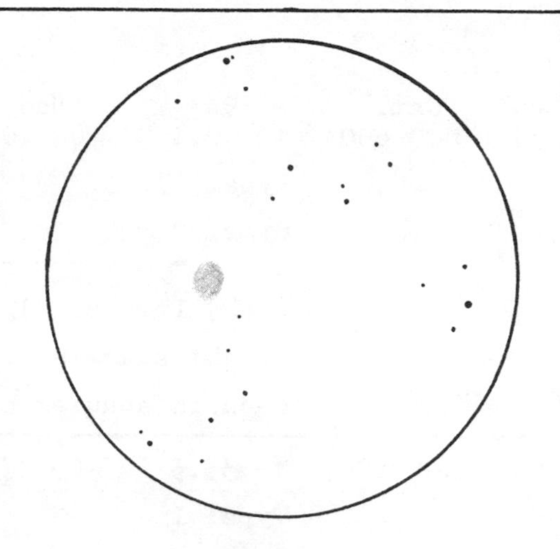

NGC 6781
$8\frac{1}{2}$-inch x102
Field 24'

E.S. Barker

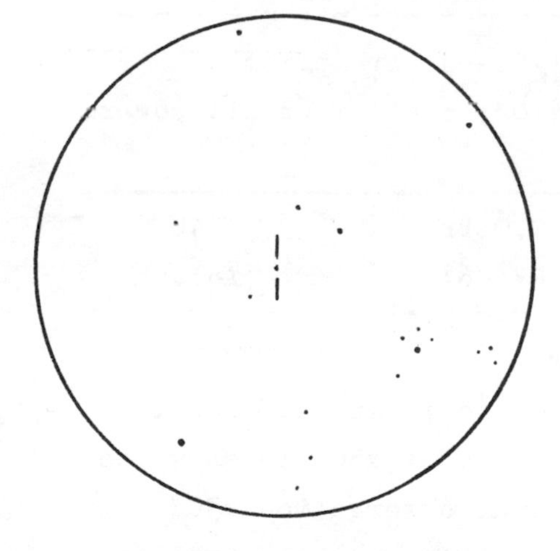

NGC 6790
$8\frac{1}{2}$-inch x102
Field 24'

E.S. Barker

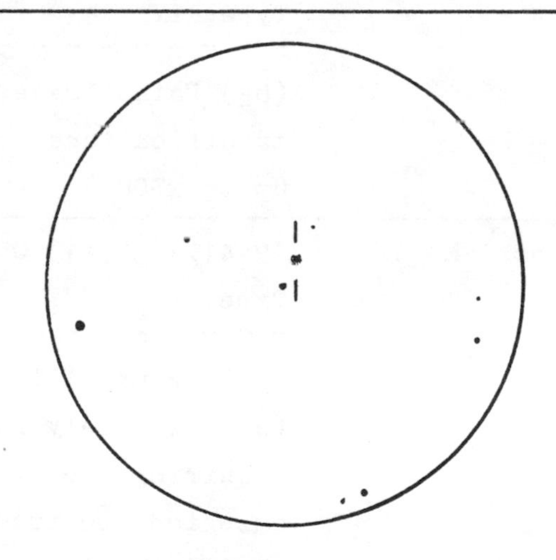

NGC 6803
$8\frac{1}{2}$-inch x204
Field 12'

E.S. Barker

WS	Cat.	RA	Dec	m^n	m^s	AD
53	NGC 6804	19 30.4	+09 10	12.2v	13.4	63 x 50
		Type: IV+II		r = 1.65		
		Star: C		RV -13		Aql.
		(16½) Irregularly round nebula in rich field; bright star in N.f. edge; x176 central star seen in annular nebulosity.				
54	NGC 6807	19 33.3	+05 37	13.8	19.3	2
		Type: I		r = 9.94		
		Star?		RV -67.7		Aql.
		(8½) Faint prism image requiring MP or HP; completely stellar on all magnifications; lies about 70" SE of a 10 mag star.				
55	Me1-1	19 38.1	+15 35	12.6	-	2
		Type: IV		RV -6		Aql.
		(8½) Faint image, only seen with prism at magnifications over x102; stellar on all powers up to x308.				
56	M1-74	19 41.1	+15 06	9.0SBr	-	10
		Type: I		RV -6		Aql.
		(100) Blue stellar image. (8½) Extremely faint object; very difficult requiring use of the prism at x308 to show the emission spectrum; normal observation x308 showed the image as possibly just non-stellar; 13 mag star about 40" to the E.				

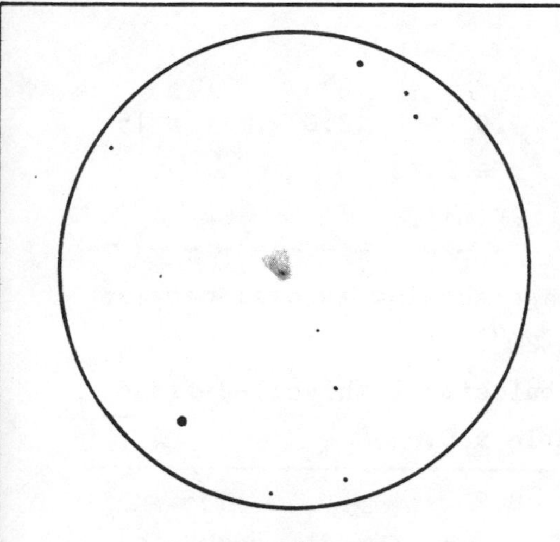

NGC 6804
8½-inch x204
Field 12'

E.S. Barker

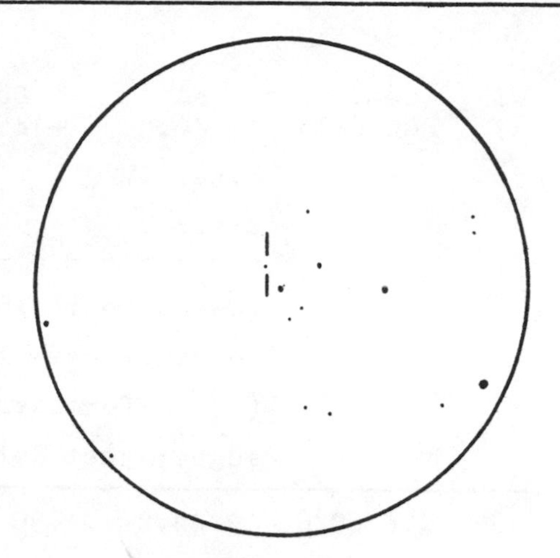

NGC 6807
8½-inch x102
Field 24'

E.S. Barker

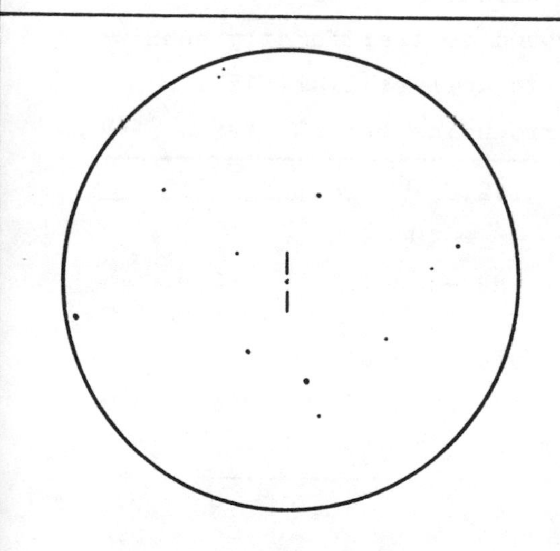

Me1-1
8½-inch x102
Field 24'

E.S. Barker

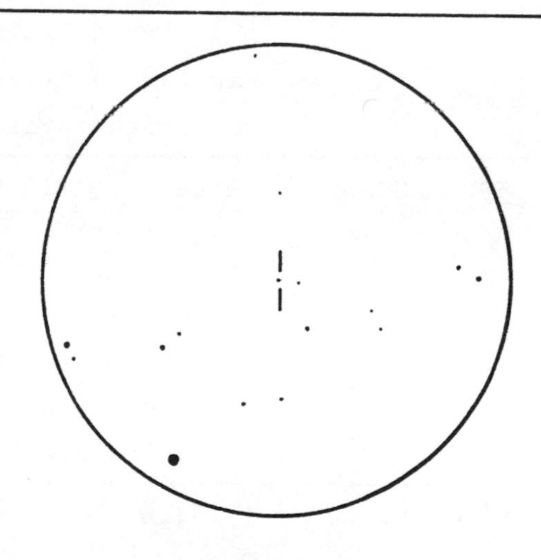

M1-74
8½-inch x204
Field 12'

E.S. Barker

WS	Cat.	RA	Dec	m^n	m^s	AD
57	NGC 6818	19 42.5	-14 12	9.9	12.0	22 x 15
		Type: IV		r = 2.38		
		Star: ?		RV -13.8		Sgr.

(8) Quite bright image showing as oval nebula; no other detail; PA 20°.

(6) Uniform oval nebulosity with well-defined edges; just detectable x60.

WS	Cat.	RA	Dec	m^n	m^s	AD
58	NGC 6826	19 44.2	+50 28	8.8	9.9	135
		Type: IIIa		r = 1.26		27 x 24
		Star: 06/WN6		RV -6.2		Cyg.

(100) 35" x 30"; major axis 130° - 310°; pale green with soft edges and bright star.

(16½) Dark curved area between star and f. edge; very bright, circular.

(8) Very bright, round centre; faintly seen markings difficult to define; diam. 15".

(6) Bright oval surrounding bright star.

WS	Cat.	RA	Dec	m^n	m^s	AD
59	NGC 6833	19 49.0	+48 54	12.0	20.3	2
		Type: I		r = 10.09		
		Star: C		RV -108.7		Cyg.

(36) Bright nebula; diam. about 4".

(8½) Faintish prism image requiring use of MP; 7' E of variable RT Cyg.

WS	Cat.	RA	Dec	m^n	m^s	AD
60	NGC 6842	19 54.0	+29 13	13.6	16.0	50 x 45
		Type: IV		r = 1.62		
		Star: ?		RV ?		Vul.

(16½) Diffuse roundish disk with ill-defined edges; x351 central star or brighter centre; x527 central star probable plus bright patch on N edge; 2 stars on f. edge; in rich field.

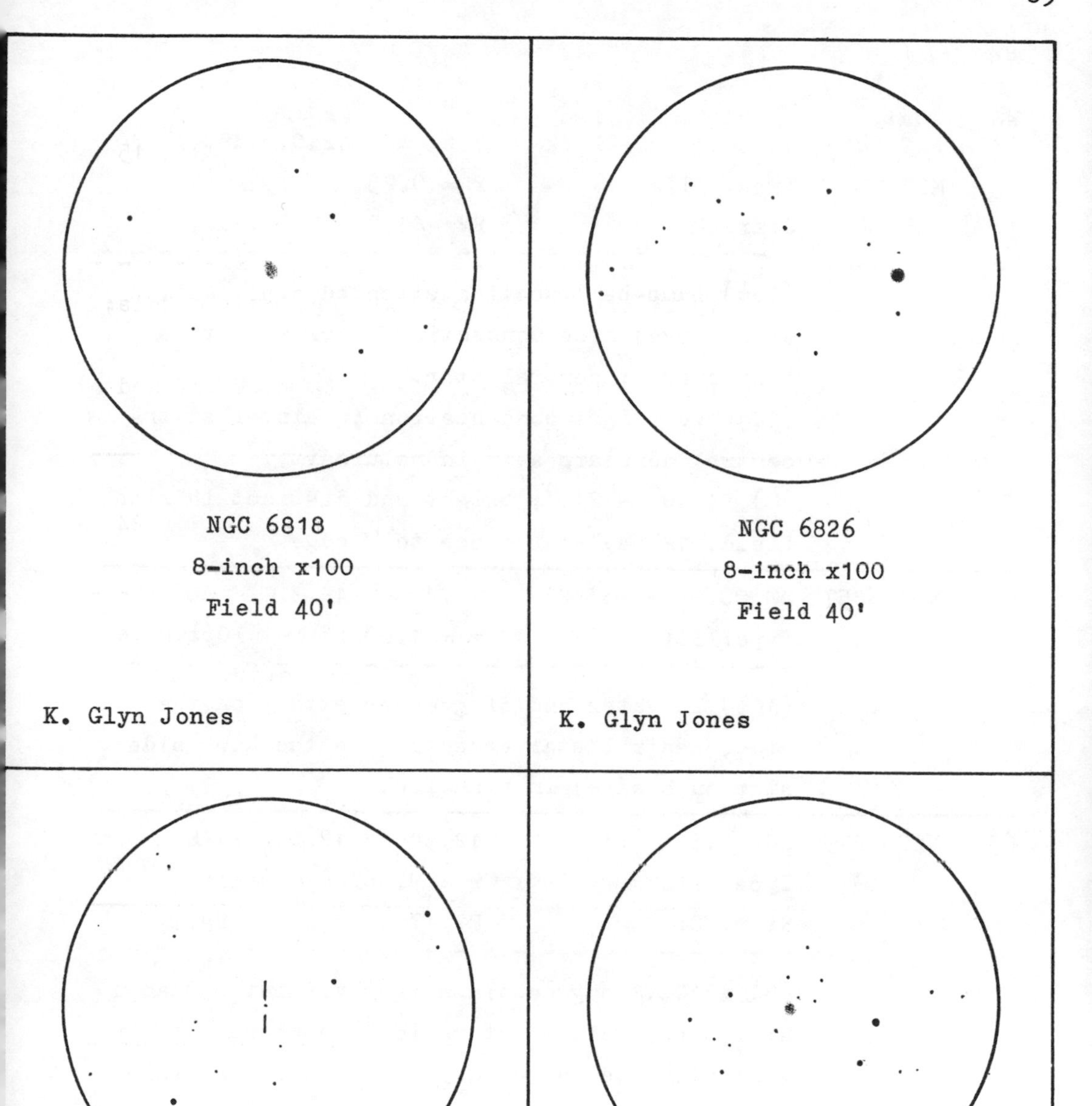

NGC 6818
8-inch x100
Field 40'

K. Glyn Jones

NGC 6826
8-inch x100
Field 40'

K. Glyn Jones

NGC 6833
8½-inch x102
Field 24'

E.S. Barker

NGC 6842
8-inch x125
Field 23'

P. Brennan

WS	Cat.	RA	Dec	m^n	m^s	AD
61	NGC 6853	19 58.5	+22 42	7.6	12.0	480 x 240
	M27	Type: IIIa		r = 0.25		
		Star: C		RV -41.5		Vul.

(16½) Dumb-bell section extended S.p., N.f. with curved arcs connecting; dark vacuities inside arcs; central star.
(10) Two bright condensations; fainter at the centre; no stars seen in nebulosity.
(8) PA 30° - 210°; bright and distinct in rich field; 12 mag star close to W edge.

WS	Cat.	RA	Dec	m^n	m^s	AD
62	NGC 6857	20 00.9	+33 27	11.4	14.3	40
		Type: III		r = 1.80		Cyg.

(16½) Brighter and of greater extent on the p. side; central star eccentric to the N.p. side; star on S edge; in rich field.

WS	Cat.	RA	Dec	m^n	m^s	AD
63	NGC 6879	20 09.3	+16 50	12.5v	15.2	8
		Type: IIa		r = 9.96		
		Star: Of		RV +7.1		Sge.

(8) x310 12 mag nebula slightly fainter than a star S.p.; fainter stars in line close p.; virtually stellar at HP; detected using prism at x125.

WS	Cat.	RA	Dec	m^n	m^s	AD
64	NGC 6884	20 09.6	+46 23	12.6	18.6	6
		Type: IIb		r = 5.82		
		Star: C		RV -35.6		Cyg.

(16½) Blue disk at x527 with faint envelope fading at the edges.
(8½) Stellar at LP; x308 fuzzy-edged disk with brighter centre; bright prism image.

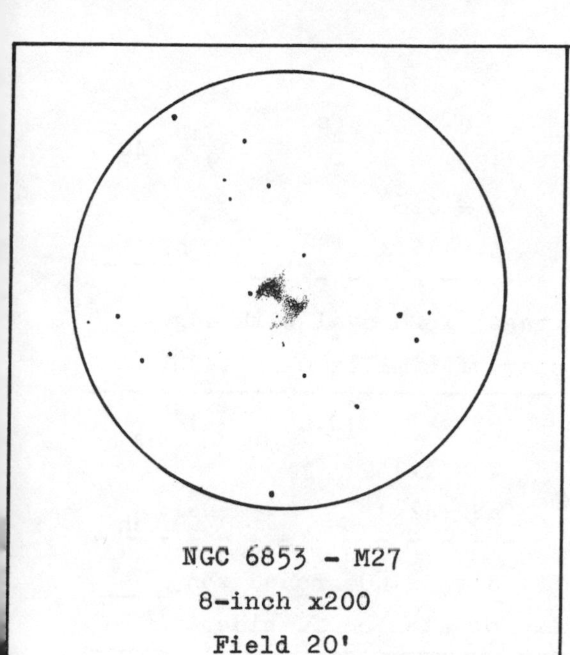

NGC 6853 - M27
8-inch x200
Field 20'

K. Glyn Jones

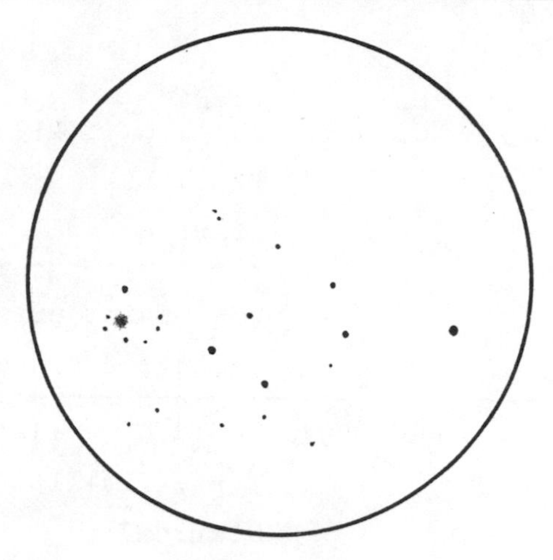

NGC 6857
8-inch x65
Field 45'

P. Brennan

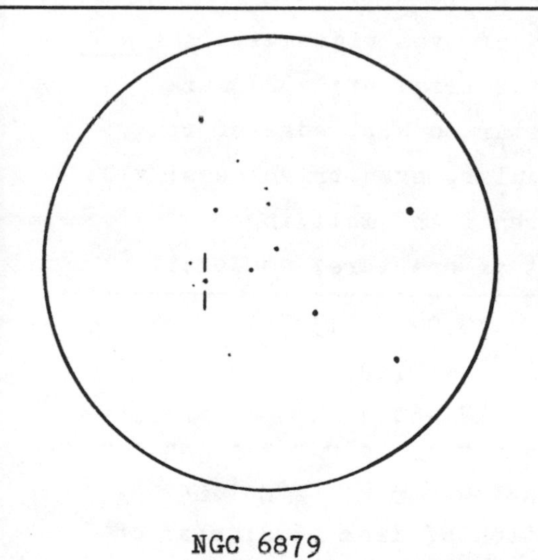

NGC 6879
8-inch x125
Field 25'

P. Brennan

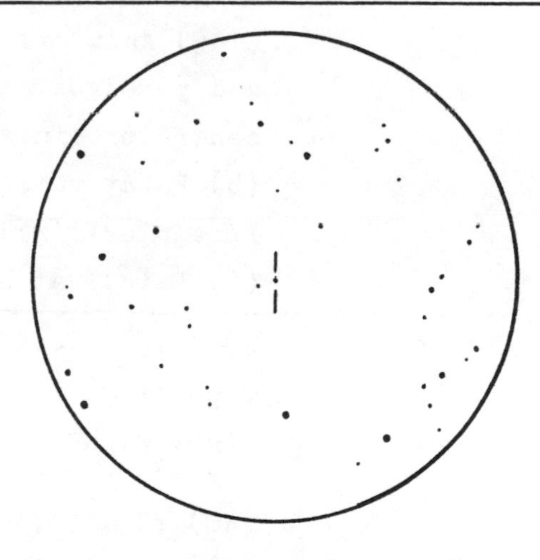

NGC 6884
$8\frac{1}{2}$-inch x51
Field 45'

E.S. Barker

WS	Cat.	RA	Dec	m^n	m^s	AD
65	NGC 6886	20 11.6	+19 55	12.2	-	9 x 6

Type: II — r = 6.67

Star: - — RV -36.4 — Sge.

(8½) Faint prism image; x308 oval with edges fading; 40" N of star of similar magnitude.

WS	Cat.	RA	Dec	m^n	m^s	AD
66	NGC 6891	20 14.0	+12 38	9.0v	10.0	12

Type: IIa+IIb — r = 3.14

Star: 07 — RV +42.1 — Del.

(16½) Stellar at LP, disk x160, round x333 with small appendage or star on f. side.

WS	Cat.	RA	Dec	m^n	m^s	AD
67	NGC 6894	20 15.4	+30 30	12.0v	17.5	57

Type: IV+II — r = 1.49

Star:? — RV -58 — Cyg.

(16½) Annular; ring of even width with the N and p. edges a little brighter; x527 dark centre prominent; star on N.p. edge of ring.
(8) Faint x61; circular, even brightness; x183 irregular circumference and mottling.
(6) Faint, at limit of aperture; no detail.

WS	Cat.	RA	Dec	m^n	m^s	AD
68	IC 4997	20 19.0	+16 40	12.0	13.7	5

Type: I — r = 8.80

Star: W7/W8 — RV -64.4 — Del.

(40) 2" x 1"; elongated E - W; bright green.
(8½) Slight indication of disk x308; star of equal mag 1' SE; bright prism image.

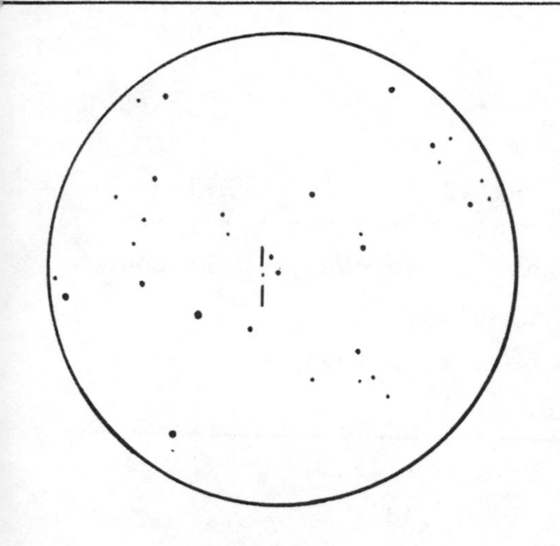

NGC 6886
$8\frac{1}{2}$-inch x51
Field 45'

E.S. Barker

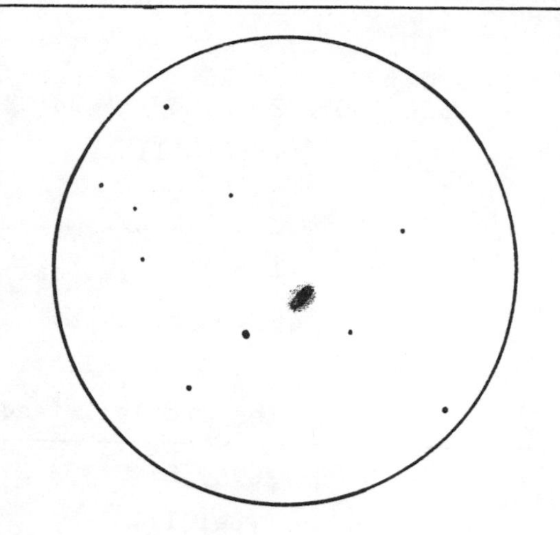

NGC 6891
18-inch x200
Field 13'

J.K. Irving

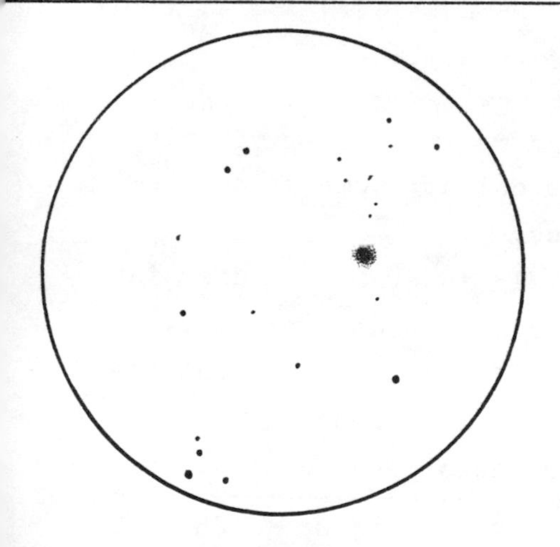

NGC 6894
8-inch x125
Field 25'

P. Brennan

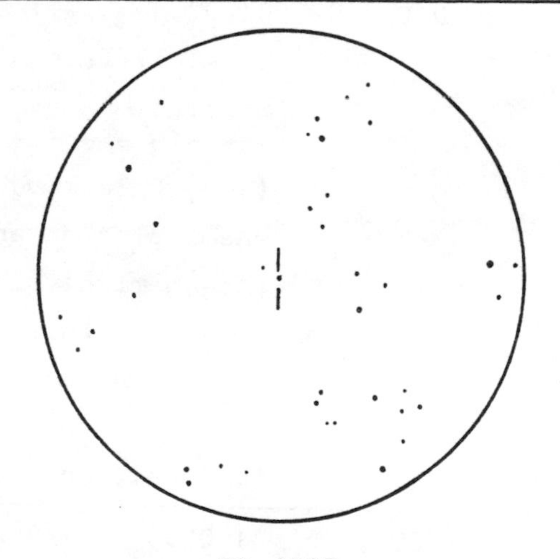

IC 4997
$8\frac{1}{2}$-inch x51
Field 45'

E.S. Barker

WS	Cat	RA	Dec	m^n	m^s	AD
69	NGC 6905	20 21.3	+20 02	10.5v	14.2	44 x 37

Type: IIIa r = 1.82

Star: WC6 RV -4 Del

(100) Elliptical; sharp edges except near ends of major axis; central star.

(8) Small, distinct, hazy ellipse; brighter in the middle; stands HP well.

WS	Cat	RA	Dec	m^n	m^s	AD
70	NGC 7008	20 59.8	+54 27	10.0	12.0	86 x 69

Type: III r = 1.14

Star: C RV -73 Cyg

(10) Mottled; star displaced to SE.

(8½) x308 partially annular; irregular in brightness; possible condensation N.f. star.

(6) Curving, fan-shape; PA 0°; N part bright.

WS	Cat	RA	Dec	m^n	m^s	AD
71	NGC 7009	21 02.9	-11 28	8.4	12.0	44 x 26

Type: IV+IIIa r = 1.63

Star: C RV -46.4 Aqr

(100) Pale turquoise colour; complex double-shell plus extensions.

(16½) Bright blue oval; inner bright disk in fainter nebula; S and S.p. sides brighter.

(8) Bright oval, PA 70° - 150°; solid outline and traces of ansae; no star seen.

(6) Oval, no star or ansae seen to x400.

WS	Cat	RA	Dec	m^n	m^s	AD
72	NGC 7026	21 05.5	+47 45	9.5	14.0	25 x 10

Type: IIIa r = 1.66

Star: WC RV -40.3 Cyg

(100) Light green; two parallel hazy lines brightest in the middle; no star.

(8½) Hazy rim around bright, extended nebula; N edge shows slight indentation at HP.

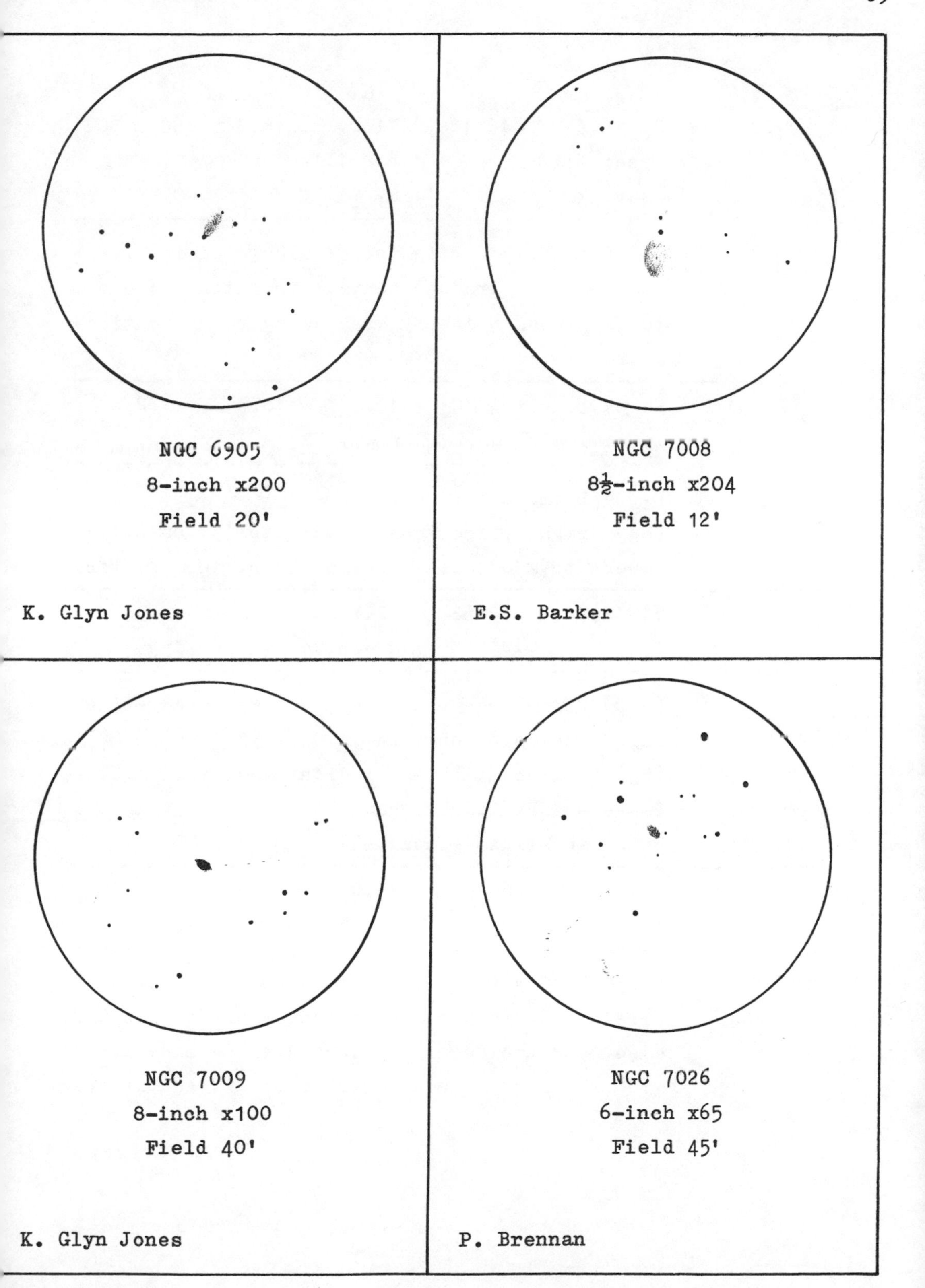
NGC 6905
8-inch x200
Field 20'
K. Glyn Jones
NGC 7008
8½-inch x204
Field 12'
E.S. Barker
NGC 7009
8-inch x100
Field 40'
K. Glyn Jones
NGC 7026
6-inch x65
Field 45'
P. Brennan

WS Cat. RA Dec m^n m^s AD

73 NGC 7048 21 13.3 +46 10 11.0v 18.3 60 x 50

Type: IIIb r = 1.19

Star: ? RV ? Cyg.

- -

(8) Low surface brightness; elongated in PA 150° - 330°; gradual central brightening; due to faintness not easy when using magnifications over x125.

---

74 IC 5117 21 31.5 +44 30 10.5 18.3 2

Type: I r = 7.78 Cyg.

- -

(100) Diam. 2"; green.

($8\frac{1}{2}$) Bright prism image x204; stellar on all powers to x308; lies 10" to 15" p. 10 mag star.

---

75 Hu1-2 21 32.2 +39 52 12.7 - 9

Type: II r = 5.24 Cyg.

- -

(100) Pale green with a shape like an egg-timer; faint central star; diam. 10" x 7".

($8\frac{1}{2}$) Stellar at LP and indications of irregular image at HP; bright enough to be found with the prism in slightly hazy sky.

---

76 NGC 7139 21 45.3 +63 32 13.0 18.0 86 x 70

Type: IIIb r = 1.21 Cep.

- -

($16\frac{1}{2}$) Pretty bright, quite large and shows as irregularly round; A star lies close to the S.f. edge, and the nebula suspected to be annular; x160 the S.p. side appears to be little brighter.

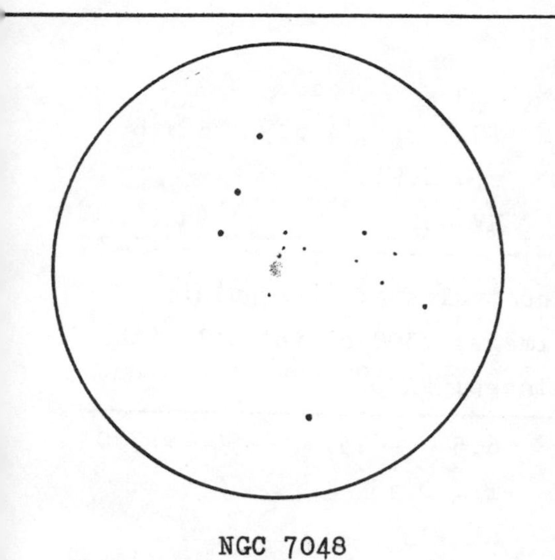

NGC 7048
8-inch x125
Field 20'

P. Brennan

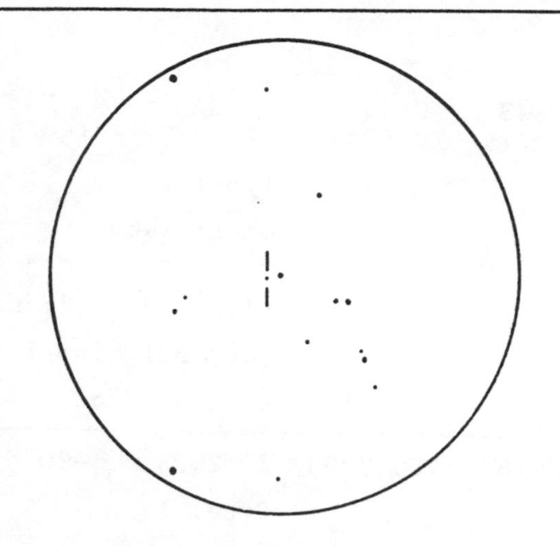

IC 5117
$8\frac{1}{2}$-inch x102
Field 24'

E.S. Barker

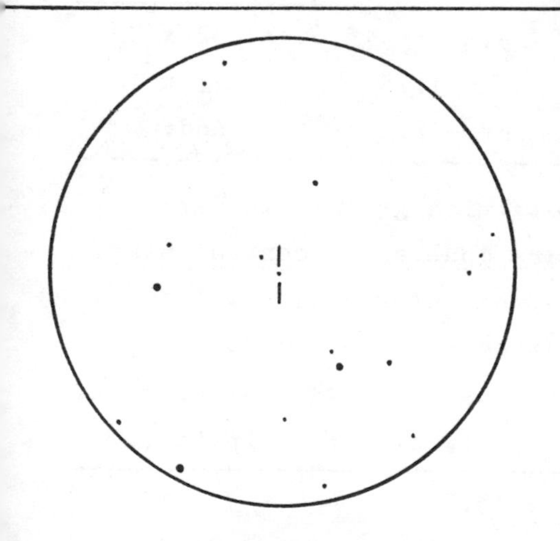

Hu1-2
$8\frac{1}{2}$-inch x102
Field 24'

E.S. Barker

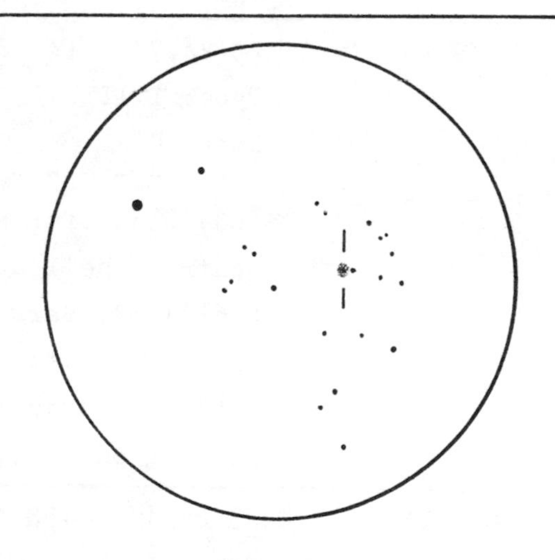

NGC 7139
6-inch x65
Field 45'

P. Brennan

WS	Cat.	RA	Dec	m^n	m^s	AD
77	IC 5217	22 22.9	+50 51	10.0	14.6	8 x 6
		Type: II		r = 4.94		
		Star: WN6		RV -98.6		Lac.

(100) 7" x 5"; no central star; irregular.
(8½) Bright prism image; x308 elliptical with even surface brightness; PA 0°.

WS	Cat.	RA	Dec	m^n	m^s	AD
78	NGC 7293	22 28.3	-20 55	6.5	13.3	900 x 720
		Type: IV		r = 0.15		
		Star: ?		RV -15		Aqr.

(16½) Annular; ring very wide and broken on the N.p. edge; brighter on the N.f. and S edges; two stars at centre.
(8) x40 very faint; irregular brightness and hints of annularity in good seeing.

WS	Cat.	RA	Dec	m^n	m^s	AD
79	NGC 7662	23 24.7	+42 24	9.4	15.6	32 x 28
		Type: IV+III		r = 1.78		17 x 14
		Star: C		RV -12.2		And.

(60) Beautiful sight; rich green with dark centre; shell appears double; no central star.
(16½) x527 dark centre; outer edge split on f. side; suspected knot on f. edge; no star.
(8½) x308 traces of ring and dark centre, ring showing on S.p., S and f. sides; elliptical.

WS	Cat.	RA	Dec	m^n	m^s	AD
80	Hb 12	23 25.1	+58 03	13.2	-	10
		Type: I		r = 1.85		Cas.

(36) Bright, possibly non-stellar.
(8½) Bright prism image; x308 possible disk of hazy appearance; 11 mag star 75" N.f. and fainter star about same distance N.p.

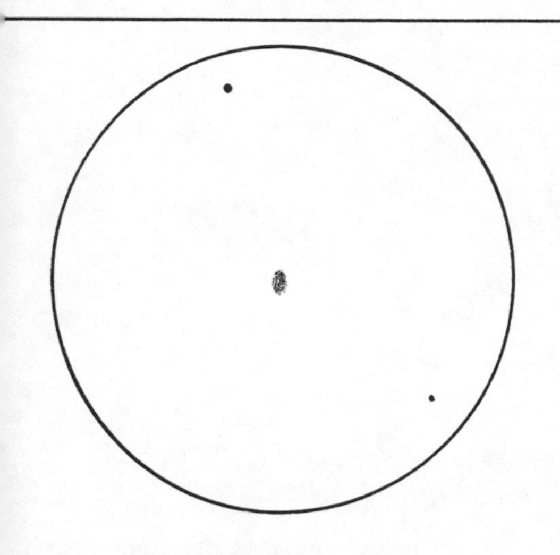

IC 5217
8½-inch x308
Field 6'

E.S. Barker

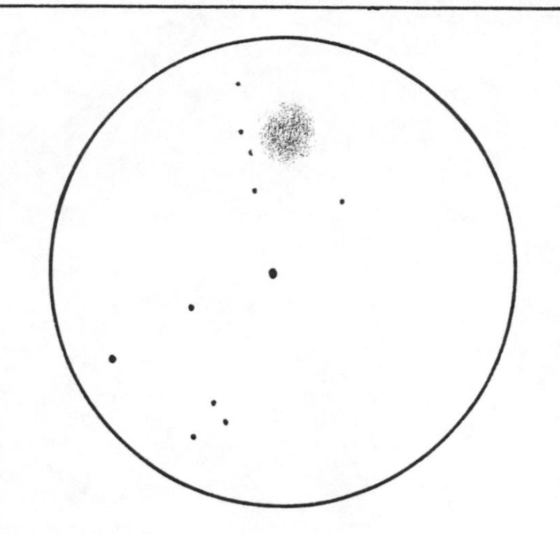

NGC 7293
6-inch x40
Field 75'

P. Brennan

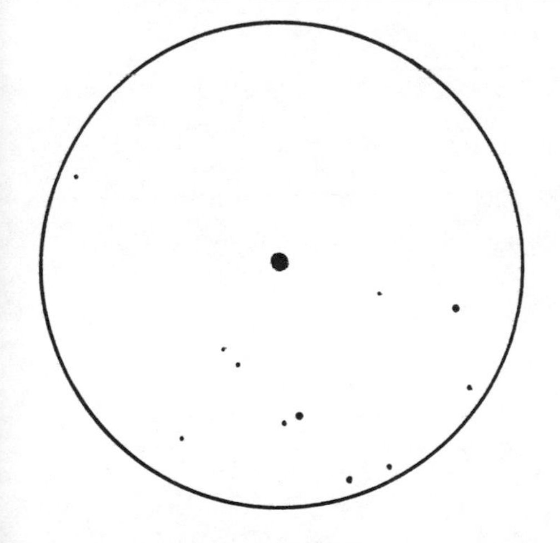

NGC 7662
8½-inch x102
Field 24'

E.S. Barker

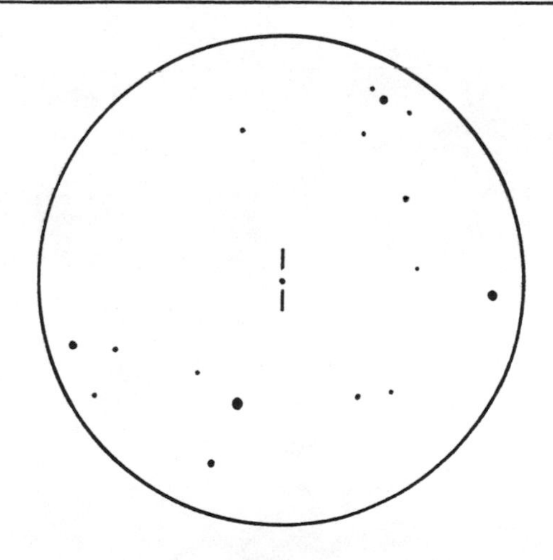

Hb 12
8½-inch x102
Field 24'

E.S. Barker

LIST OF ADDITIONAL OBJECTS.

Nebula.	RA	Dec	m^n	m^s	AD	Type.
M3-2	07 13.9	-27 45	9SBr		9	IIIb
M1-16	07 36.1	-19 24	11SBr		3	VI
NGC 6369	17 27.8	-23 45	10.4	16.6	28	IV+II
M1-22	17 33.6	-18 53	10SBr		9 x 8	IV
M1-23	17 35.8	-18 45	10SBr		8 x 6	II
M1-24	17 36.7	-19 36	9SBr		7 x 5	II
M2-15	17 45.4	-16 17	10SBr		6 x 5	II
M2-17	17 50.6	-17 53	9SBr		7 x 5	II
Vy1-2	17 53.3	+28 00	7.5SBr		5 x 4	II
Hb 6	17 53.7	-21 44	13.0		5	II
M1-33	17 57.5	-15 32	7SBr		5 x 4	II
NGC 6537	18 03.8	-19 51	12.5		5	IIa
M1-43	18 10.3	-18 47	9SBr		6 x 4	II
NGC 6565	18 10.3	-28 11	13.2		10 x 8	IV
NGC 6578	18 14.8	-20 27	14.0	15.8	8	IIa
Cn3-1	18 16.7	+10 08	11.8		6	II
M2-40	18 20.0	-06 02	9SBr		5 x 4	II
M2-39	18 20.5	-24 11	8SBr		3	II
M2-42	18 21.0	-24 10	8SBr		4 x 3	II
NGC 6620	18 21.4	-26 50	15.0	15.8	5	IIb
M3-53	18 22.7	-11 07	10SBr		6 x 3	II
H1-67	18 23.5	-22 35	9SBr		6 x 5	II
NGC 6629	18 24.2	-23 13	10.6	13.6	16 x 14	IIa
M1-46	18 26.4	-15 33	8SBr		11 x 10	IV+II
M1-47	18 27.7	-21 48	8SBr		4 x 4	II
M1-48	18 28.0	-19 06	7SBr		10	II
M1-49	18 28.8	-13 55	10SBr		10 x 2	VI
NGC 6644	18 31.0	-25 09	12.2		2	I
M1-50	18 31.8	-18 17	9SBr		10	II
M1-52	18 32.5	-14 53	10SBr		7 x 6	II
IC 4732	18 32.5	-22 38	13.3		2	I
M1-53	18 34.3	-17 37	8SBr		6 x 5	II

List of Additional Objects.

Nebula	RA	Dec	m^n	m^s	AD	Type.
M1-54	18 34.6	-17 01	9SBr		16 x 10	III
M2-44	18 36.3	-03 07	9SBr		8 x 7	IV
M1-57	18 39.0	-10 41	9SBr		9 x 7	II
M1-58	18 41.6	-11 07	9SBr		7 x 5	II
M2-46	18 45.1	-08 29	9SBr		5 x 3	III
M1-62	18 48.9	-22 35	7SBr		3	II
M1-63	18 50.1	-13 12	8SBr		4	II
M1-65	18 55.5	+00 50	8SBr		4 x 3	II
M1-75	20 03.8	+31 22	11SBr		16 x 11	IIIb
M1-77	21 18.2	+46 13	8SBr		7	IIa
M2-50	21 56.8	+51 34	9SBr		4 x 3	II
M2-52	22 19.6	+57 29	11SBr		12 x 11	III.

PART FOUR : A CATALOGUE OF GASEOUS NEBULAE.

INTRODUCTION.

The catalogue contains observations of 93 nebulae. As in many cases individually numbered nebulae are part of a large complex (e.g., NGC 6960, 6979, 6992-5 comprise the Cygnus Loop) the Webb Society catalogue numbers down the left hand sides of the pages will total 81. Full distribution of data is as follows.

The left sections of the pages show the Webb Society catalogue number (WS) followed by the actual designation of the nebula (NGC, IC etc.). In a few cases there will be found objects to which no designations have been assigned.

The upperline of each entry, opposite the catalogue number, shows the following data.

(a) Positions for 1975.0

(b) Magnitudes of the exciting stars (m^s). Where the nebula is connected with a star cluster, the word 'cluster' will be found (e.g., NGC 1624, IC 410).

(c) Approximate angular diameter of the nebula (AD). The sources for these are the Atlas Coeli Catalogue, the catalogues of Sharpless and Lynds supplemented by inspection of the Palomar Sky Survey prints.

(d) Types of nebulae. (E) refers to emission nebulae and (R) to reflection nebulae. In some cases the letters (E/R) will be found, these being examples of either mixed nebulae, part emission, part reflection (due to a star of intermediate temperature) or those reflecting emission-line stars.

(e) The abbreviated form of the relevant constellation.

No distances are shown, in many cases these not having been determined, while in other cases the figures are not very reliable.

In some cases additional data will be found below the above-mentioned information.

Visual Observations.

Data below the dotted line covers the observations of the nebulae, the format being identical to that of the catalogue of planetary nebulae.

Field Drawings.

Eleven drawings have been selected, covering a variety of nebulae. In some cases whole field drawings are shown, in others only the immediate area surrounding the object. In most cases scales in either arcminutes or arcseconds are shown. The drawings will be found towards the end of the catalogue.

A Catalogue of Gaseous Nebulae.

Following the catalogue will be found a supplementary list of possible diffuse nebulae for further observation.

List of Observers.

The following list shows the names of the observers whose work features in the catalogue, plus details of their locations and respective telescopes, all reflectors unless shown otherwise.

Observer	Telescope	Location
D.A. Allen	100-inch	Mount Wilson, U.S.A.
	60	" " "
	98	R.G.O. Herstmonceux, U.K.
	60	Teneriffe, Canary Islands.
	50	Mount Stromlo, Australia.
	40	Chile.
	40	Sutherland, S. Africa.
	20	" "
	12	Coonabarabran, Australia.
	12 o.g.	Cambridge, U.K.
	8 o.g.	" "
M.J. Thomson	$16\frac{1}{2}$	Santa Barbara, U.S.A.
J. Perkins	14, 10	Kirkby-in-Ashfield, U.K.
G. Hurst	10	Earls Barton, U.K.
E.S. Barker	$8\frac{1}{2}$, 16 x 50	Herne Bay, U.K.
S.J. Hynes	$8\frac{1}{2}$, 10 x 50	Wistaston, "
C. Nugent	$8\frac{1}{2}$	Upton, U.K.
P. Brennan	8	Regina, Canada.
S. Selleck,	8	Santa Barbara, U.S.A.
I. Genner,	8, 3	Newcastle-upon-Tyne, U.K.
K. Sturdy	6, 4	Helmsley, U.K.
G. Gough	$2\frac{1}{2}$ o.g.	Cambridge, "
P.A. Bennett	$1\frac{1}{4}$ o.g.	Dorking, U.K.

The following observers are all members of the Albireo Club in Hungary.

Observer	Telescope	Observer	Telescope
A. Kökény	$11\frac{3}{4}$	J. Papp	6
B. Szentmártoni	6 RFT, $4\frac{1}{2}$ RFT	G. Gombás	$4\frac{1}{4}$ RFT, $3\frac{1}{4}$, 2 o.g.
	3 o.g.	V.G. Kiszel	4
L. Baracskai	6	T. Juhász	$1\frac{3}{4}$
P. Brlás	6 RFT	Z. Szoboszlai	10 x 80
J. Dankó	6	S. Tóth	10 x 80
S. Kezsthelyi	6		
G. Mohácsi	6		
Z. Hevesi	6		

Observers of Individual Nebulae.

The listing overleaf shows all the gaseous nebulae which comprise the catalogue in order of RA. As in the introduction to the previous catalogue, observers are listed by initials only.

Object	Observers
NGC 281	DAA, MJT.
1333	DAA, MJT.
1432	DAA.
1435	DAA, KS, PB, GG, ZH.
1491	ESB, KS.
41 Tau	SS.
NGC 1554-5	DAA.
1579	DAA, IG, MJT.
1624	MJT.
1788	DAA, ESB, BS, MJT, GG, ZH, JP.
IC 410	BS.
417	DAA, ESB.
NGC 1931	DAA, ESB, GH, SS, KS, MJT.
1952	DAA, SS, KS, MJT.
1973-5-7	DAA, SJH, BS, PB, JD, ZH, JP.
1976	DAA, PAB, GG, GH, JP, SS, KS.
1980	DAA, PB, JD,JP.
1982	DAA, GG, GH, SS, KS.
1990	SS.
1999	DAA, BS, MJT.
1985	MJT.
IC 431	DAA.
432	DAA, ESB.
434	PB, ZH, AK.
NGC 2023	DAA, SJH.
2024	DAA, ESB, PB, SS, KS, MJT.
IC 435	DAA, ESB.
NGC 2067-8	DAA, SJH, KS, MJT.
2071	KS, MJT.
2149	MJT.
Ced 62	DAA.
NGC 2170	DAA, MJT.
Ced 66	DAA.
NGC 2174-5	BS, MJT, ZH, JP.
2182	MJT.
2183	DAA, MJT.
2185	DAA, MJT.
2237-8-9 2246	MJT, AK.
2245	DAA, IG, MJT.
2247	DAA, MJT.
2261	DAA, ESB, BS, MJT.
2316	MJT.
NGC 2327	MJT.
IC 466	GH.
NGC 2359	BS, MJT, GG.
VY CMa	DAA.
NGC 2467	DAA, MJT.
IC 4592	ESB.
IC 4601	ESB.
NGC 6302	DAA, MJT.
6514	DAA, ESB, KS.
6523	DAA, ESB, JP, KS, MJT.
6559	MJT.
6589	MJT.
6590	MJT.
6611	KS, MJT.
IC 1287	MJT.
Sh 61	DAA.
M 1-92	DAA, CN, KS, LB, TJ, AK, GM.
NGC 6813	MJT.
6823	DAA.
6888	SJH, MJT.
6910	PB, ZH, AK.
6913	PB, JD.
6914	DAA, SJH, MJT.
	PB, ZH.
44 Cyg	PB, ZH, JP.
IC 5067	DAA, PB, GG, ZH, VGK, ZS, ST.
NGC 6960-92-95	DAA, IG, GH, CN, SJH, KS, BS, PB, MJT, GG, ZH, JP.
IC 5076	MJT, ZH, JP.
NGC 7000	DAA, SJH, BS, PB, GG, ZH, VGK, ZS, ST.
7023	DAA, MJT.
7027	DAA, JP, MJT.
IC 1369	ZH.
NGC 7076	MJT.
7129	DAA, ESB, MJT.
IC 5146	GH.
NGC 7538	GH.
2315 +60	DAA.
NGC 7635	MJT.

WS	Cat	RA	Dec	m^s	AD	T	Con
81	NGC 281	00 51.7	+56 29	7.9	25 x 20	E	Cas

- -

(60) Distinct milky area around the multiple star Burnham 1; dark absorption lane to the S very prominent, producing a sharp edge to the nebula; at least 5' diam.

(16½) Large, easy object in a rich field; very diffuse.

(8) Sky around Bu 1 is clearly bright; the absorption edge is still distinct.

WS	Cat	RA	Dec	m^s	AD	T	Con
82	NGC 1333	03 27.6	+31 17	10.8	9 x 5	E/R	Per

- -

(60) Not an easy object unless the sky is very dark; irregular, elongated nebula.

(16½) Boomerang-shaped, curving southwards from the bright star; faint star or nebulous knot on S.p. edge.

WS	Cat	RA	Dec	m^s	AD	T	Con
83	NGC 1432 1435	03 44.9	+23 53		30 x 20	R	Tau

Cluster (Pleiades). NGC 1432 is 20' in AD and surrounds 20 Tau. NGC 1435 = IC 349 (as in Atlas Coeli) surrounds 23 Tau. Fainter nebulae surround other Pleiads.

- -

(12) Both nebulae easily seen; 1432 is fainter and circular; 1435 is comet-shaped and stretches southwards from Merope.

(6) Both nebulae are distinct but appear smaller than in (12).

(4) 1435 is not difficult under good conditions but usually requires averted vision.

WS	Cat	RA	Dec	m^s	AD	T	Con
84	NGC 1491	04 01.4	+51 14	11.0	3	E	Per

- -

(8½) Bright object, especially to the W of the star; visible even in a hazy sky.
(6) Faint; extends mostly SW of the star.

* * *

WS	Cat	RA	Dec	m^s	AD	T	Con
85		04 05.0	+27 32	5.3 (41 Tau)			Tau

- -

No nebula is apparent on Palomar prints. However, Selleck reports a convincing observation with an 8-inch. This object needs further study.

* * *

WS	Cat	RA	Dec	m^s	AD	T	Con
86	NGC 1554 1555	04 20.4	+19 28	var.	5	R	Tau

Celebrated variable nebula around T Tau. Once fairly easy, currently very difficult.

- -

(60) Difficult object, only visible by using averted vision with T Tau out of the field; a curved arc about the star, almost lost near its mid point, hence the two NGC numbers.

* * *

WS	Cat	RA	Dec	m^s	AD	T	Con
87	NGC 1579	04 28.5	+35 13	17	12 x 8	R	Per

Has an emission spectrum because it reflects an emission-line star (Lk Hα 101). The star is reddened by about 10 magnitudes and therefore does not hinder observation.

- -

(100, 60) Almost circular nebula, brightest on S.f. edge, where it is abruptly cut off; does not extend as far as illuminating star which lies further south.
(16½) Large, bright, irregular; brighter at the f. end.
(12) Irregular, fairly bright; brightest in the centre; diam. 6'; estimated magnitude 11.

WS	Cat	RA	Dec	m^s	AD	T	Con
88	NGC 1624	04 38.4	+50 24	Cluster	5	E	Per

(16½) Large, irregular nebula, fairly uniform in brightness and containing five or six stars; probably elongated N.p., S.f.

WS	Cat	RA	Dec	m^s	AD	T	Con
89	NGC 1788	05 05.8	-03 22	10.0	8 x 5	E/R	Ori

(60) Bright, irregular elliptical nebula elongated in PA 140° - 320°.

(16½) Easily seen; several stars involved, brightest to SE; a close double lies N.p.

(8½) Bright, easy object visible even in a hazy sky.

(6) Elongated, faint nebulosity.

(3) Seen by averted vision.

WS	Cat	RA	Dec	m^s	AD	T	Con
90	IC 410	05 22.1	+33 29	Cluster	20	E	Aur

The coordinates refer to the brightest portion N.p. the cluster NGC 1893.

(6) Seen by averted vision and possibly directly.

WS	Cat	RA	Dec	m^s	AD	T	Con
91	IC 417	05 27.5	+34 22	Cluster	20	E	Aur

The coordinates refer to the brightest portion 8' S.f. phi Aur.

(100) A rather faint nebula brightest around a multiple star.

(8½) Detailed, irregular nebula found to SE of NGC 1931; involves a triangle of 9 to 10 magnitude stars.

WS   Cat   RA   Dec m^s AD   T   Con

92   NGC 1931   05 29.8   +34 14   Cluster   4   E   Aur

This is the brightest portion of a large area of nebulosity including IC 410 and 417

---

(60) Bright, complex nebula involving a miniature "Trapezium" of four stars with fainter stars to the south.

($16\frac{1}{2}$) Bright nebula in a rich field; three stars of the Trapezium seen.

(10, $8\frac{1}{2}$, 8) Elongated in PA 30°; high power an advantage; five stars visible within.

(6) Central 2' visible; 3 stars seen at HP.

---

93   NGC 1952   05 33.0   +22 00   15(pulsed)   6 x 4   SNR   Tau

M1

The 'Crab' Nebula: nebular remnant of supernova of 1054 AD. The stellar remnant is a pulsar rotating 30 times a second and flashing twice at every rotation at all wavelengths from gamma-rays to radio. Discovery of M1 by Messier prompted his catalogue.

($16\frac{1}{2}$) Bright, irregular nebula; brightest along N - S axis and brighter f. than p. this; one dark bay seen at f. end.

(12) Unusual nebulosity because of its irregularity and lack of a central star or central brightening; several bays suspected scalloped from the outer edges.

(6) Irregular; oval outline well seen.

---

94   NGC 1973   05 34.0   -04 47   4.7   40 x 25   E   Ori

1975

1977

A large emission nebula with three bright centres around exciting stars. NGC 1977 surrounds 42 Ori, the 4.7 mag star.

(12, 10) The entire group easily seen to

WS	Cat	RA	Dec	m^s	AD	T	Con
94 cont.							

be a common nebula; NGC 1977 is the hardest due to its brighter star.

(8, 6) 1073 and 1975 seen as separate nebulae, the former lying to the S and appearing larger; 1975 appears greenish in the centre; 1977 requires averted vision.

($3\frac{1}{2}$) 1973 and 1975 visible but difficult.

WS	Cat	RA	Dec	m^s	AD	T	Con
95	NGC 1976 M42	05 34.0	-05 24	Cluster	(600)	E	Ori

This is the centre of the nebulosity which covers the whole of the Orion constellation. It contains the Trapezium (theta 1 Ori) which ionizes many of the outlying nebulae as well as 1976 itself. There are dense neutral molecular clouds behind M42, and curling part way in front to produce the dark bay. Numerous young stars and infrared sources are associated with this object.

- -

M42 is the brightest H II region in the northern skies; it is easily visible to the naked eye and has probably been observed by every owner of a telescope or binoculars. Observations contributed to this programme have been made with a range of instruments from 60-inch down. All observers record similar detail, and to list these or give a drawing would be superfluous. The large apertures reveal more detail of the nebula, but suffer from a restricted field. The best view is probably had with an 8 to 10-inch or a larger RFT. Two observers noted colour, one describing the striking contrast between the green inner regions around the Trapezium and the red outer streaks (12-inch).

WS	Cat	RA	Dec	m^s	AD	T	Con
96	NGC 1980	05 34.2	-05 55	2.9	14	E	Ori

Nebula around iota Ori.

- -

(12) Quite difficult nebula because of the bright star involved.

(6) Averted vision usually neccessary; about 5'.

WS	Cat	RA	Dec	m^s	AD	T	Con
97	NGC 1982 M43	05 34.3	-05 17	9.1	5	E	Ori

A locally brighter portion of M42 appearing distinct in small telescopes.

- -

Like M42 observed by most people. In the majority of reports the nebula is seen to be fairly regular regular and roughly circular, but a very large aperture (60) reveals much structure.

WS	Cat	RA	Dec	m^s	AD	T	Con
98	NGC 1990	05 35.0	-01 13	1.8	30	E/R	Ori

- -

This nebula, though bright, is very hard to see and requires extremely good seeing. Only Selleck (8) recorded a convincing sighting, and negative reports were received from apertures up to (12).

WS	Cat	RA	Dec	m^s	AD	T	Con
99	NGC 1999	05 35.3	-06 44	9-10	12	E/R	Ori

Exciting star V380 Ori.

- -

(60) Easy, bright nebula, almost circular but with a pronounced absorption bay.

(16½, 12) The absorption appears as a dark bay p. the star.

(6) Suspected.

WS	Cat	RA	Dec	m^s	AD	T	Con
100	NGC 1985	05 36.1	+31 59	13	1	E	Aur

- -

(16½) Small, faint; requires high power.

WS	Cat	RA	Dec	m^s	AD	T	Con
101	IC 431	05 39.0	-01 28	7.8	5 x 3	R	Ori

- -

(60) Oval nebula, elongated in PA $30^\circ - 210^\circ$; bright central star.

WS	Cat	RA	Dec	m^s	AD	T	Con
102	IC 432	05 39.8	-01 30	7.1	6 x 4	R	Ori

- -

(60) Twin of IC 431; only slightly brighter.
($8\frac{1}{2}$) Small nebula; quite obvious.

WS	Cat	RA	Dec	m^s	AD	T	Con
103	IC 434	05 39.9	-02 51	1.9	60 x 10	E	Ori

This nebula stretches south from its exciting star and contains the dark 'Horsehead' Nebula. It was sought, but not seen, by many observers.

- -

(12) Very difficult, but most of the nebula seen.
(6) Best seen near zeta Ori by averted vision; a suspicion of its southern extension was reported by two observers.

WS	Cat	RA	Dec	m^s	AD	T	Con
104	NGC 2023	05 40.5	-02 14	7.8	10	E/R	Ori

- -

(60) Easily seen as a large, circular nebula fading outwards from its central star.
(12) A conspicuous object about 5' diam.
(8) Difficult object; perhaps 4' diam.

WS	Cat	RA	Dec	m^s	AD	T	Con
105	NGC 2024	05 40.8	-01 51	-	30	E	Ori

The exciting star or stars are hidden behind a central dust cloud which permits only the outer portions to be seen.

- -

($16\frac{1}{2}$) Large and bright; in the form of a Greek pi with the crosspiece to the SE; all patches appear mottled at HP; three stars within.
(12) Circular; central square absorption patch.
($8\frac{1}{2}$, 8, 6) Well seen with no apparent loss of detail over (12).

WS	Cat	RA	Dec	m^s	AD	T	Con
106	IC 435	05 41.8	-02 19	8.2	3	R	Ori

- -

(60) Bright, easy nebula.

(12) Somewhat more difficult than neighbouring NGC 2023; 3' diam.

(8½) Easily seen; takes high magnification well; some structure suspected, notably absorption to the E.

WS	Cat	RA	Dec	m^s	AD	T	Con
107	NGC 2067 2068 M78	05 45.5	+00 17	10.3+10.6	6 x 3	R	Ori

The two nebulae are connected, or at best separated by a weak absorption lane.
NGC 2068 is brighter; 2067 lies 4' N.p.

- -

(60) Seen only in a bright sky; NGC 2068 is bright and elongated and contains two almost equal stars; 2067 suspected; other fainter stars visible.

(16½) 2068 appears mottled at HP, dusky areas lying between the two bright stars and near the E end; 2067 very diffuse and seen easily only by averted vision.

(12, 8, 6) Only 2068 seen, stretching southwards from a pair of stars.

WS	Cat	RA	Dec	m^s	AD	T	Con
108	NGC 2071	05 45.9	+00 17	10.4	3	R	Ori

This is part of the same nebular complex as 2067-8, but is more distinctly separated.

- -

(16½) Large, irregular nebula surrounding a double star; extends further to the S of this.

(12, 6) Rather difficult; roughly circular.

WS	Cat	RA	Dec	m^s	AD	T	Con
109	NGC 2149	06 02.3	-09 44	-	3	R	Mon

- -

(16½) Very faint nebula surrounding a star; better seen at high power.

WS	Cat	RA	Dec	m^s	AD	T	Con
110	Ced 62	06 06.3	+18 42	13	2 x 1	R	Ori

This nebula is illuminated by a faint emission line star (Lk Hα 208) which is surrounded by a dense equatorial dust cloud: a planetary system in the making. The dust allows the light to escape only through the poles. We view the star equator-on, so it thus appears dimmed by dust.

- -

(60) Quite bright; en egg-timer nebula that appears as two symmetrical fans of light projecting N and S from the star.

111	NGC 2170	06 06.4	-06 23	10.2	2	R	Mon

This is the most westerly of the Mon R2 association of reflection nebulae. Near NGC 2170 is a highly obscured H II region which appears to be connected with the association. This is probably a system like M42, but with the dust cloud between us and the H II region. The Mon R2 association comprises the stars on our side of the dark cloud bright enough to illuminate it.

- -

(50) Bright, easy object; elongated roughly E - W and about 2' x 1'.5.

(16½) Bright; extends further S.p. the star.

112	Ced 66	06 06.9	-08 18	9.1(Mon R2)		R	Mon

- -

(50) Faint nebula around a rather bright star.

113	NGC 2174 2175	06 08.2	+20 34	7.4	30 x 25	E	Ori

- -

(16½) A faint nebula surrounds the bright star; a second nebula to the north is also faint and

WS	Cat	RA	Dec	m^s	AD	T	Con

113 cont. elongated N.p., S.f.; this contains three fainter stars.

(6) The cluster is very prominent; the nebula appears faintly as a background much larger than the cluster and brightest north of the main star.

(3) Only the brightest portion (NGC 2174) is visible.

114	NGC 2182	06 08.3	-06 19	9.1 (Mon R2)	3	R	Mon

(16½) Almost circular nebula surrounding a bright star; possibly more extended south.

115	NGC 2183	06 09.6	-06 12	13.5 (Mon R2)	1.5	R	Mon

(50) A bright, elongated nebula surrounding a bright star; unusually easy to see for a reflection nebula.

(16½) Bright, irregular nebula; perhaps wedge-shaped.

116	NGC 2185	06 09.9	-06 12	12.5 (Mon R2)	2	R	Mon

(50) An easy, bright nebula; a fainter nebula surrounds a star S.p. this.

(16½) Faint; two of four stars S.p. may lie in the nebula.

117	NGC 2237	06 29.7	+04 54	Cluster	60	E	Mon

2238 2239 2246

The 'Rosette' Nebula around the cluster NGC 2244. The entire nebula is large, the NGC entries being portions of it.

(16½) NGC 2246 noted as a large, faint irregular nebula in a rich field.

WS	Cat	RA	Dec	m^s	AD	T	Con
117 cont.							

(12) Seen as a few isolated sections well away from the distracting cluster; insufficient information is given to identify which NGC numbers were seen.

118	NGC 2245	06 31.3	+10 11	10.8	5 x 3	R	Mon

(60) Bright, irregular nebula 2' to 3' across and containing a group of stars.
(16½) Appears fan-shaped; the apex to the N contains the bright star.
(8½) Comet-like nebula with apex to the NE; 1' to 2' in diam.

119	NGC 2247	06 31.7	+10 22	8.5	4 x 3	R	Mon

(60) Extensive area of nebulosity around the brightest star; quite easy.
(16½) Difficult; slightly more extended S.p. the star.

120	NGC 2261	06 37.8	+08 45	var(R Mon)	5 x 3	R	Mon

R Mon is the tip of this, which is Hubble's variable nebula. It is unclear whether R Mon is really stellar or the brightest part of this nebula. Magnitude of R Mon about 11.

(100) A magnificent sweeping fan fading to the N of R Mon and curving gently as if fanned by a light breeze; f. edge of the nebula more defined.
(16½) The p. edge of the fan weakened by a dark bay just N of R Mon; at HP a second nebulous knot seen just W of R Mon.
(8) Fan-shape evident; p. edge seems fainter.
(3) Barely visible to averted vision; no shape.

WS	Cat	RA	Dec	m^s	AD	T	Con
121	NGC 2264	06 39.6	+09 47	Cluster	50 x 20	E	Mon

(100, 60) The brightest portion of the nebulosity, in a group of stars south of S Mon, is quite bright. It is irregular and about 4' across; The 'Cone' Nebula to the S is visible as a bright rim in the 100-inch only.

WS	Cat	RA	Dec	m^s	AD	T	Con
122	NGC 2316	06 58.5	-07 04	-	4 x 3	R/E	Mon

(16½) Bright nebula lying north of three stars; a faint double star at centre, the components almost equal.

WS	Cat	RA	Dec	m^s	AD	T	Con
123	NGC 2327	07 03.1	-11 16	10.0	20	R	CMa

Lies among faint emission nebulae.

(16½) Irregular, small nebula surrounding a double star; brightest p. the northern of the two stars.

WS	Cat	RA	Dec	m^s	AD	T	Con
124	IC 466	07 05.6	-04 17	11.5	1.0	E	Mon

(10) Small, round nebula lying between two 10 mag stars separated N - S; easily seen at low magnification.

WS	Cat	RA	Dec	m^s	AD	T	Con
125	NGC 2359	07 16.6	-13 09	Cluster	8 x 6	E	CMa

IC 468 is the fainter NW portion of this object.

(16½) The entire nebula visible, NGC 2359 appearing as a bright wedge along the southern edge; many stars involved.

(3) About 10 stars in an easy, circular nebula.

(2) Quite conspicuous.

WS	Cat	RA	Dec	m^s	AD	T	Con
126	VY CMa	07 21.9	-25 43	var.	0.3 x 0.2	R	CMa

This nebula contains a number of small knots which have often been mistaken for stars. VY CMa therefore appears in double star catalogues, (Innes and others). Attempts at orbital solutions failed, and it was shown by Herbig that all the knots are moving outwards from the central star. Curiously, for such an object, VY CMa is a very cool star. Its normal mag is 7.5, and it is a very bright infrared source. The nebula was discovered by Guerin in 1917 using a 7-inch meridian circle.

- -

(100) The nebula is very easy and appears to be extended mostly to the W. The star is extremely red; several of the star-like knots are visible.

WS	Cat	RA	Dec	m^s	AD	T	Con
127	NGC 2467	07 52.3	-26 20	8.5	4	E	Pup

- -

(40) A beautiful, complex nebula 4' across and containing several stars; the brightest is at the northern edge.

(16½) The centre appears darker, and the outer ring is brightest S. and S.p.

WS	Cat	RA	Dec	m^s	AD	T	Con
128	IC 4592	16 10.6	-19 24	4.3	160 x 45	R	Sco

- -

(8½) Nebula seen very faintly around nu Sco and its companion, extending furthest to the SW and NE; possibly a dark lane to the SE.

WS	Cat	RA	Dec	m^s	AD	T	Con
129	IC 4601	16 17.4	-20 10	7.0	20 x 10	R	Sco

- -

(8½) Probably seen around the single and two double stars involved.

WS	Cat	RA	Dec	m^s	AD	T	Con
130	NGC 6302	17 12.2	-37 05	16.0	4 x 1.5	E	Sco

This nebula is included, despite its far southern declination, because it is a bright and unusual object.

- -

(40) One of the most striking of diffuse nebulae; the central region is a bright, vividly green ellipse elongated N - S and measuring about 25" x 12"; around this is a fainter, irregular nebula stretching mostly E - W in a series of bright jets, curves and blobs; the effect is one of an explosive event having occured in the centre.

(16½) The central part is very bright and the p. extension appears the brighter.

WS	Cat	RA	Dec	m^s	AD	T	Con
131	NGC 6514 M20	18 00.4	-23 02	6.9	30 x 25	E+R	Sgr

The 'Trifid' Nebula. Most of this is emission nebulosity but an area of reflection nebula lies to the north.

- -

(100, 60) Surprisingly faint for a Messier object; it is large and milky, and does not seem to brighten very much towards the centre; three prominent dark lanes, almost evenly spaced 120° apart, radiate from the central multiple star (hence the name).

(12) A difficult object at the altitude it reaches in Britain; the radial dark lanes are barely suspected.

(8½, 6) Faint, hazy nebula, seen to extend N to a second bright star.

(16 x 50) Much easier than it appears in larger apertures.

WS Cat RA Dec m^s AD T Con

132 NGC 6523 18 02.3 -24 22 5.9 55 x 35 E Sgr

M8

The 'Lagoon' Nebula. The brightest portion is often called the 'Hourglass' from its shape.

- -

(100, 60) The 'Hourglass' extremely prominent and sharply defined; it dominates a large, faint nebula which overfills the field.

(20) Nebulosity is visible over a 40' area; a dark lane crosses the nebula.

(16½, 12) Over 30' across; the dark lane curves across from NW to SE and is quite wide; the W section of the nebula is mottled and the 'Hourglass' seen as a bright nebular arc.

(8½, 6) Only the brightest portion visible except under perfect conditions, when the dark lane can be seen to be like a bay that intercepts the nebula.

---

133 NGC 6559 18 08.3 -24 08 9.8 8 x 5 E Sgr

The declination is wrongly quoted in the Atlas Coeli Catalogue.

- -

(16½) Large and fairly bright; a curved arc, broader and brighter at the N.p. end, where two stars are invested, the brighter a double.

---

134 NGC 6589 18 14.9 -19 48 9.5 5 x 3 R Sgr

This and NGC 6590 (overleaf) are the brightest portions of an extensive area of reflection nebulae. IC 1283-4 is a slightly fainter portion which is included in the list of additional nebulae following this catalogue.

- -

(16½) Two stars in a faint, irregular nebula.

WS	Cat	RA	Dec	m^s	AD	T	Con
135	NGC 6590	18 15.0	-19 53	10.0	3 x 2	R	Sgr

(16½) A bright oval nebula containing two stars of equal magnitude.

WS	Cat	RA	Dec	m^s	AD	T	Con
136	NGC 6611 M16	18 17.6	-13 47	Cluster	35 x 30	E	Sgr

(16½) A mottled, rectangular nebula which divides the cluster almost through the centre and partly obscures the fainter stars.
(6) Just seen east of the bright double which lies in the greatest density of stars.

WS	Cat	RA	Dec	m^s	AD	T	Con
137	NGC 6618 M17	18 19.4	-16 11	?	50 x 40	E	Sgr

The dimensions refer to the very faint outer nebula. There is a prominent dark molecular cloud in M17, and the ionizing star or stars may be hidden behind this. The A0 star listed in Atlas Coeli is a foreground object.

(60) The full omega shape clearly visible, though the western bar is very faint; the eastern bar is very mottled and streaked; faint nebulosity is seen over a large field.
(12, 8½) The details of the main bar are well seen (see drawing); the curve of nebulosity to the W encloses a very dark obscured region.
(6) The details of the bar not readily seen, but the bar and curve are still easy objects.

WS	Cat	RA	Dec	m^s	AD	T	Con
138	IC 1287	18 30.1	-10 49	5.8	45 x 35	R	Sct

The 5.8 mag star is STF 2325.

(16½) A difficult nebula faintly seen immediately around the bright double star and probably more extensive to the south.

WS	Cat	RA	Dec	m^s	AD	T	Con
139	Sh 61	18 31.6	-05 02	13.0	2	R	Sct

- -

(100) A faint nebula surrounding a pair of red stars 20" apart; diam. about 1' x 1'.5.

WS	Cat	RA	Dec	m^s	AD	T	Con
140	M1-92	19 35.3	+29 29	20.0	0.2 x 0.1	R	Cyg

The 'Footprint' Nebula. The two blobs reflect an emission-line star which is hidden in the dark dust lane which separates them. Discovered by Minkowski and included in an appendix to his first list of planetary nebulae.

- -

(100, 60) Estimated 8" x 3"; the brighter blob (the sole) is about 4" x 3"; at PA 140° lies a fainter portion (the heel) about 3" x 3" and slightly irregular in shape; 4" S of the sole is a small spot of light which may be a star or another part of the nebula; the surface brightness is very high; no sign of the illuminating star.

(12, 8, 6) Under high power the nebulous nature can just be seen; most observers were able to suspect the heel and give the correct PA; at low powers looks like an 11 mag star.

(2) Appears entirely stellar.

WS	Cat	RA	Dec	m^s	AD	T	Con
141	NGC 6813	19 39.2	+27 14	-	3	E	Sge

- -

(16½) Fairly bright, irregular nebulosity surrounding two stars of which the S.f. (the brighter) is a double.

WS	Cat	RA	Dec	m^s	AD	T	Con
142	NGC 6823	19 42.2	+23 15	Cluster	40 x 30	E	Vul

- -

(8) Two fairly prominent nebulous patches in a field possibly nebulous throughout; 4 stars involved in the N patch and 2 in the S.

WS	Cat	RA	Dec	m^s	AD	T	Con
143	NGC 6888	20 21.6	+38 21	7.4	18 x 8	E(SNR?)	Cyg

- -

(16½) A long filamentary nebula; four stars lie in a diamond shape and NGC 6888 curves from the E star through the N star to end midway between the S and W stars; brightest just W of the southernmost star where it appears cometary.
(8½) Only the brightest portion is visible; in the field of RS Cyg.

WS	Cat	RA	Dec	m^s	AD	T	Con
144		20 22.2	+40 42	Cluster	80 x 50	E	Cyg

This is the brightest of a number of nebulae that surround gamma Cyg, and is best seen near the cluster NGC 6910.

- -

(6) The nebula is faintly seen among the stars of the cluster; best with averted vision.

WS	Cat	RA	Dec	m^s	AD	T	Con
145		20 23.1	+38 26	Cluster	-	-	Cyg

- -

Although not apparent on Palomar photographs, nebulosity was suspected by two observers using 6-inch RFT's in the vicinity of M29.

WS	Cat	RA	Dec	m^s	AD	T	Con
146	NGC 6914	20 23.9	+42 13	9.0	13 x 12	R	Cyg

This is the brightest of three reflection nebulae. 6914a and b lie at $20^h\ 23^m.9$ $+42^\circ\ 07'$ and $23^h\ 23^m.4$ $+42^\circ\ 03'$. The whole field is weakly nebulous.

- -

(100) The three nebulae visible; none very large or detailed and nebulosity is suspected over the entite field.
(8½) Nebulosity around a double star (6914a).
(6) 6914 faintly seen; possible PA of 80°.

WS	Cat	RA	Dec	m^s	AD	T	Con
147		20 30.0	+36 51	6.3	7	R	Cyg

- -

(6) Faintly seen, mostly by averted vision, to surround 44 Cyg and a nearby pair of fainter stars.

148	IC 5067	20 47.1	+44 16	6.0	85 x 75	E	Cyg

This is the brightest part of the 'Pelican' Nebula, and the brightest diffuse nebula in the IC. The other portion, IC 5070, is not usually seen. It is unclear which star is responsible for its excitation, but alpha Cyg is certainly not, as suggested in Atlas Coeli. Probably the 6 mag O star.

- -

(12) With low power this nebula can be extracted from the plethora of stars that surround it; the "neck" of the 'Pelican' can be made out fairly clearly.

(6, 4) The larger field makes the nebula stand out better; up to 30' of it can be seen.

(3) Difficult with averted vision; 10 x 80 binoculars show it better.

149	NGC 6960 6979 6992 6995	20 49.5	+30 50	-	210	SNR	Cyg

The 'Veil' or 'Cirrus' Nebula. Coordinates refer to the centre of the entire nebula, but the brightest portions are at $20^h\ 44^m.6\ +30^\circ\ 35'$ (6960) and $20^h\ 55^m.3\ +31^\circ\ 36'$ (6992-5). 6979 was not reported by any observer.

- -

($16\frac{1}{2}$) The long, filamentary nature easily seen; in places, especially just N of 42 Cyg, it is split into two by a dark lane; the section S of 52 Cyg is broader.

WS Cat RA Dec m^s AD T Con

149 cont. (14, 12, 10, 8, 6) The section to the N of 52 Cyg quite easily visible though readily lost in a bright sky; the S portion much more difficult and usually requires 52 Cyg to be placed out of the field.

(3) Difficult, but certain with averted vision.

NGC 6992-5 (16½) Bright, long curving nebula; a great deal of detail seen, especially at the wider south end.

(12, 10, 8, 6) Seen to be longer and broader than 6960, and reported to be more detailed; however, it is definitely fainter and harder to see.

(3) Easier to see than 6960.

---

150 IC 5076 20 55.0 +47 19 5.8 9 x 6 R Cyg

- -

(16½) Averted vision needed to see it well; faint, irregular and stretches further to the N and W of the star.

(6) Requires averted vision; oval.

---

151 NGC 7000 20 57.9 +44 14 6.0 120 x 100 E Cyg

The 'North America' Nebula. The largest and brightest nebula in Cyg. The exciting star is uncertain, but is certainly not alpha Cyg. Contains many young pre-main-sequence stars. A naked eye sighting was reported by Allen.

- -

(12, 6) Too big an object for this size of telescope, but because of its sharply defined edge, portions could be seen.

(4½, 4, 3) More easily recognised, though the brighter edges still an advantage.

(10 x 80, 10 x 50, 15 x 50) Well seen.

WS	Cat	RA	Dec	m^s	AD	T	Con
152	NGC 7023	21 01.7	+68 04	7.2	18	R	Cep

(60) Brightest around the star but extends a long way south in a lozenge shape; bisected by a dark lane running NE - SW and wider at its SW end.

(16½) Extends well to the south of the star; a dark bay seen.

(8) Rather faintly seen; no absorption noted.

WS	Cat	RA	Dec	m^s	AD	T	Con
153		21 02.4	+50 07		3 x 2	R	Cyg
		21 02.9	+50 10		2 x 2	R	
		21 03.1	+50 02		3 x 3		

These three reflection nebulae lie somewhat S of the positions plotted in Atlas Coeli.

(8½) Two of the nebulae seen, about 15' S of the bright star.

WS	Cat	RA	Dec	m^s	AD	T	Con
154	NGC 7027	21 06.0	+42 08	-	0.4 x 0.3	E	Cyg

Discovered by Webb in 1879. Often classified as a planetary nebula, but its optical spectrum, with highly ionized atoms and lack of a central star, make it unique. Radio maps show a symmetrical elliptical nebula, but the foreground obscuration produces an irregular appearance in the visible.

(98) Appears embryonic; the brightest portion is NW, and below the centre a dark lane almost bisects the nebula.

(14) Appears as two nebulae in contact, the eastern being annular.

(6) Small, requiring high magnification; only the brightest portion visible; PA 115°.

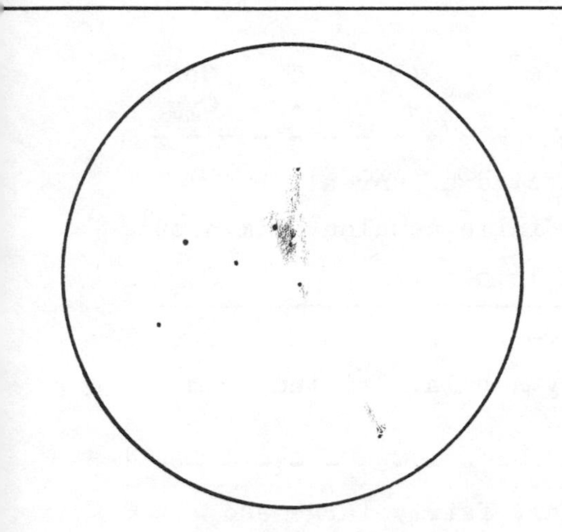

IC 417
8½-inch x102
Field 24'

E.S. Barker

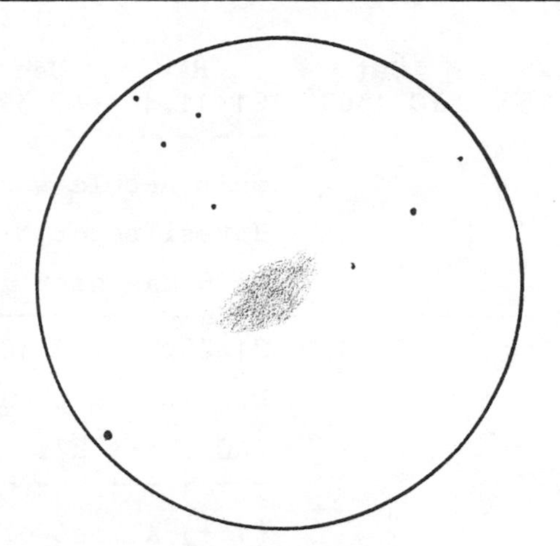

NGC 1952 - M1
6-inch x183
Field 18'

S. Selleck

5'
NGC 1973 + 1975

Based on observations by
S.J. Hynes 8½-inch
D.A. Allen 6-inch

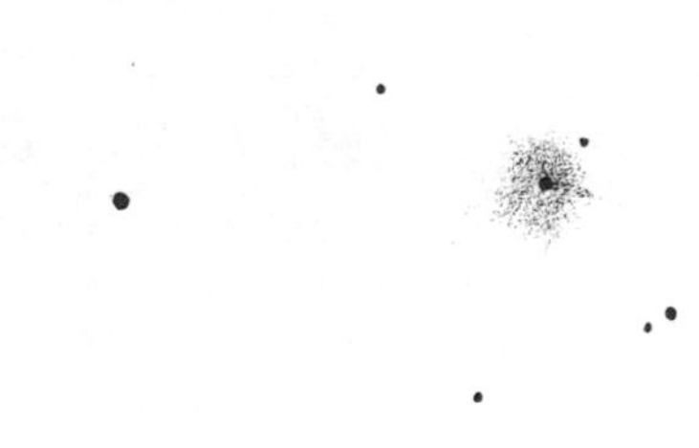

NGC 2023
8½-inch x77, x154

S.J. Hynes

WS	Cat	RA	Dec	m^s	AD	T	Con
155	IC 1369	21 11.4	+47 39	-	-	-	Cyg

This nebula was rejected as unreal, but Hevesi reports a definite nebulous rim around an 8 mag star using 15.5 cm

---

WS	Cat	RA	Dec	m^s	AD	T	Con
156	NGC 7076	21 25.2	+62 40	-	2	E	Cep

Possibly a planetary nebula. Omitted from SAO chart 27.

(16½) Almost circular, fairly large and somewhat irregular; no central star.

WS	Cat	RA	Dec	m^s	AD	T	Con
157	NGC 7129	21 42.6	+65 59	Cluster	7	R	Cep

(60) A group of six stars in a bright nebula; the four brightest stars form a trapezium of which the northern is the faintest but lies within the brightest portion of the nebula; the nebula also locally brighter near the other three stars.

(16½) Five stars recorded; nebula rather square in shape.

(8½, 8) Two stars very obvious; two others more difficult; nebulosity bright.

WS	Cat	RA	Dec	m^s	AD	T	Con
158	IC 5146	21 52.5	+47 09	Cluster	14 x 10	E	Cyg

The 'Cocoon' Nebula.

(10) Faint nebula, 12' x 7', within a sparse cluster of stars of magnitudes 10 - 13.

WS	Cat	RA	Dec	m^s	AD	T	Con
159	NGC 7538	23 12.5	+61 22	10.0	10 x 5	E	Cep

(10) Very faint, small oval nebula that surrounds two 13 magnitude stars.

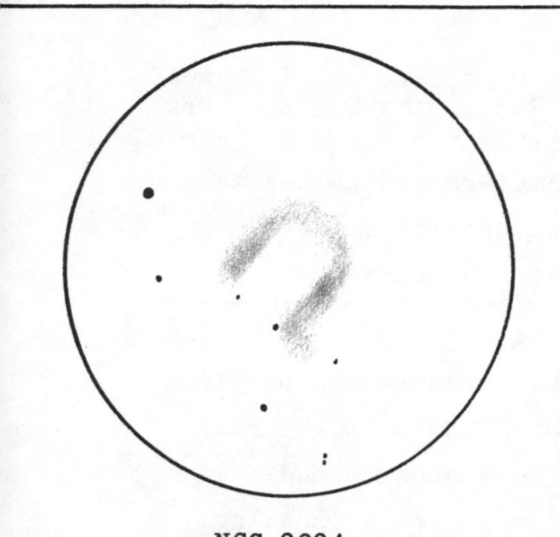

NGC 2024
8-inch x65
Field 44'

P. Brennan

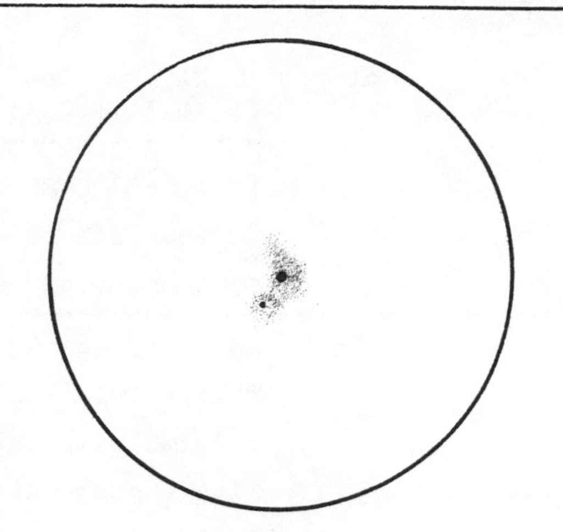

IC 4592
8½-inch x204
Field 12'

E.S. Barker

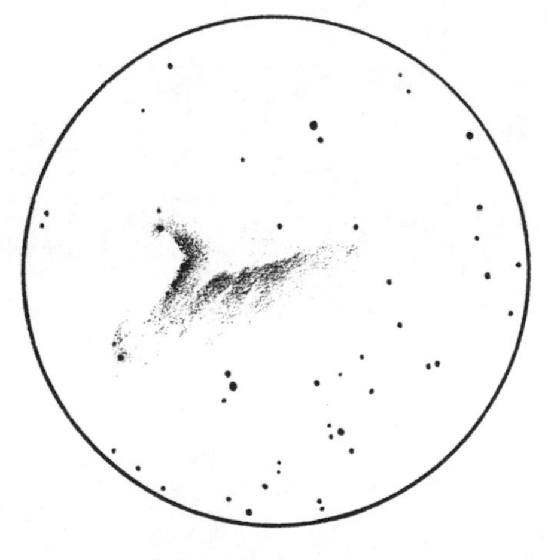

NGC 6618 - M17
8½-inch x102
Field 24'

E.S. Barker

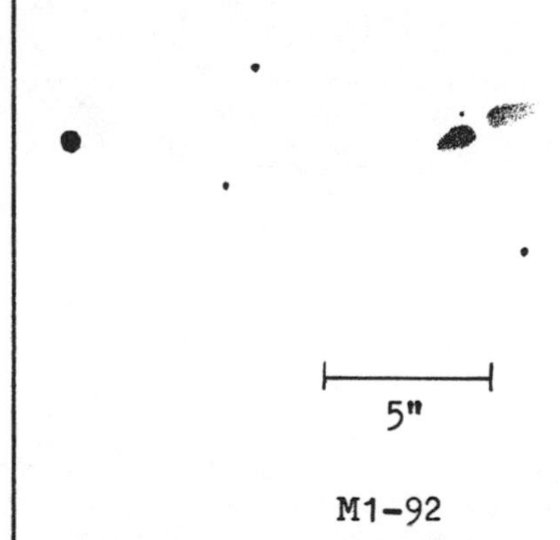

M1-92
100-inch

D.A. Allen

WS	Cat	RA	Dec	m^s	AD	T	Con
160		23 16.3	+60 43	12.5	1 x 0.5	E	Cas

(100, 60) Small, comet-shaped nebula that surrounds a red star and its fainter companion; extends to the east.

161	NGC 7635	23 19.6	+61 02	8.5	4 x 3	E	Cas

This nebula contains a remarkable bubble of neutral material which has a bright rim, like a soap bubble in a steam cloud. Included as a planetary nebula in Atlas Coeli, but this is erroneous.

(16½) A fan-shaped diffuse nebula west of a bright star; brighter on the west edge.

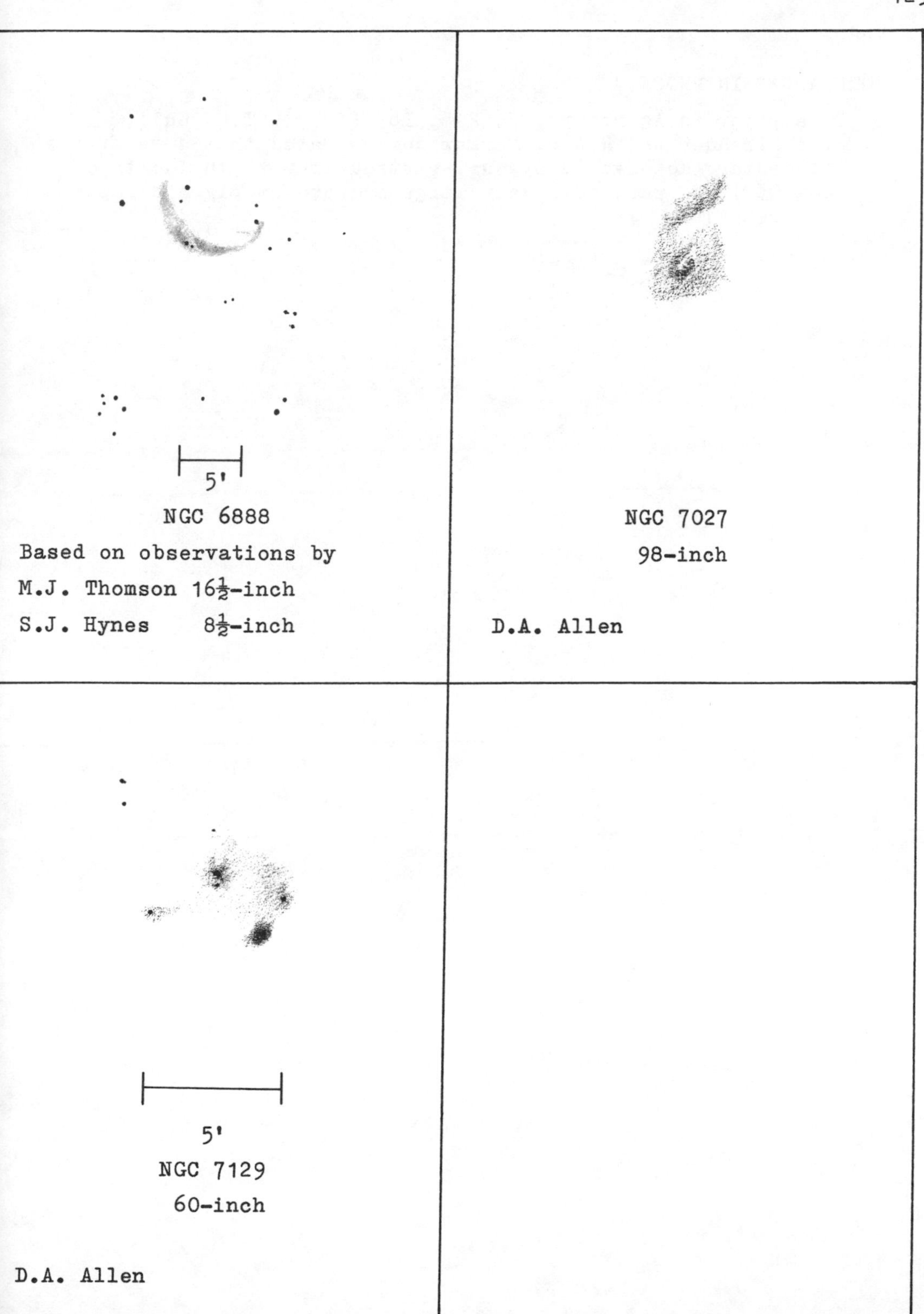

NGC 6888

Based on observations by
M.J. Thomson $16\frac{1}{2}$-inch
S.J. Hynes $8\frac{1}{2}$-inch

NGC 7027
98-inch

D.A. Allen

NGC 7129
60-inch

D.A. Allen

NOTE ADDED IN PROOF.

In a paper in Astrophys. J. 215, L69,(1977), T.R. Gull, R.P. Kirshner and R.A.R. Parker demonstrated that some of the filamentary nebulae in Cygnus, particularly to the north of beta Cyg, are part of a very large and presumably old supernova remnant.

LIST OF ADDITIONAL OBJECTS.

Nebula	RA	Dec	m^s	AD	T
NGC 7822	00 01.3	+67 02	Cluster	180	E
	00 09.3	+58 38		5	R
Sh 188	01 28.6	+58 15		9	E
NGC 896	02 23.6	+61 51	Cluster	30 x 15	E
	This is the brightest portion of a large nebula IC 1795 - IC 1805.				
IC 1848	02 49.3	+60 19	Cluster	120 x 60	E
	A large nebula; the f. end is brighter.				
IC 348	03 43.6	+32 04	5.0	120 x 60	R
	The 5 mag star is omicron Per. Two bright sections of nebulosity, that south of omicron Per is the brighter. Erroneously listed as IC 8 in Atlas Coeli Catalogue.				
NGC 1499	03 57.3	+35 43	4.1	140 x 40	E
	'California' Nebula. The 4.1 mag star is xi Per.				
Sh 288	05 11.9	+37 26		8	E
IC 405	05 14.7	+34 18	5.8	30 x 20	E
	The 5.8 mag star is AE Aur.				
Sh 263	05 20.4	+08 24	5.7	40 x 30	R/E
	05 30.8	+12 13	8.9	90 x 45	R
	Part of the faint lamda Ori emission nebula, but locally very much brighter 8' N.f. a star marked in Atlas Coeli.				
IC 426	05 35.6	-00 15	9.3	5	R
	05 37.8	+04 07	4.5	7	R
	The 4.5 mag star is omega Ori.				
Sh 231	05 38.6	+35 54		10 x 5	E
NGC 2064	05 45.1	-00 01	10.4?	4 x 2	R
	Part of the NGC 2067-68-71 complex.				
Ced 61	06 02.9	+30 29	10.5	7	R
	06 03.2	-09 45			R
Sh 241	06 03.5	+30 15		4	E

List of Additional Objects.

Nebula	RA	Dec	m^s	AD	T
Ced 71	06 10.6	-06 09	10.7	1	R
IC 2162	06 11.8	+17 59	10.0	2	E
IC 444	06 19.0	+23 18	7.0	5	R
	The 7 mag star is 12 Gem.				
NGC 2282	06 45.6	+01 21	9.7	3	R
Sh 297	07 03.3	-12 18	8.5	7	E
IC 2177	07 04.3	-10 31	7.1	20	E
	Three nebulous stars lie a little f. this object. Another bright section lies 9' north.				
Sh 301	07 08.1	-18 26	10.5	8	E
IC 4606	16 27.8	-26 22	6.5	85 x 20	E
	The 6.5 mag star is alpha Sco B, around which the nebula is brightest, and not around Antares A, as often thought.				
IC 4605	16 28.7	-25 06	4.9	60 x 40	R
	The 4.9 mag star is 22 Sco.				
IC 1274-5	18 08.2	-23 51	7.4	20 x 10	E
	IC 1274 is the fainter portion 10' north.				
IC 1283-4	18 16.1	-19 42	7.6	16 x 15	E
Sh 54	18 16.9	-11 37	Cluster	140	E
	The cluster is NGC 6604.				
IC 4706	18 18.3	-16 01	9.8	6 x 4	E
	A portion of M17 separated by a dust lane.				
Sh 71	19 00.8	+02 07	Cluster	3	E
Sh 80	19 10.4	+16 48		2	E
NGC 6820	19 41.6	+23 02	Cluster	1	E
Sh 88	19 45.4	+25 06		25	E
	Two small bright emission nebulae lie within this large faint nebula.				
Sh 93	19 54.0	+27 11		1	E
Sh 95	19 54.5	+29 14		1	E

List of Additional Objects.

Nebula	RA	Dec	m^s	AD	T
Sh 101	19 59.2	+35 12	9.5	15 x 8	E
IC 4594	20 03.8	+29 10	11.8	25	E

There are three bright components.

Nebula	RA	Dec	m^s	AD	T
IC 1318	20 15.6	+41 44	2.3		E/R
	20 15.9	+43 33			
	20 27.6	+39 52			

The 2.3 mag star is gamma Cyg. IC 1318 is a vast collection of emission nebulae that surround gamma Cyg. The coordinates refer to three bright portions, but no dimensions are given because the entire nebula covers more than 5° in filaments.

Nebula	RA	Dec	m^s	AD	T
Sh 112	20 33.0	+45 34	8.8	9 x 7	E
IV Zw 67	21 01.3	+36 36		0.5 x 0.2	E

Originally catalogued by Zwicky as a pair of compact galaxies, this reflection nebula resembles M1-92. It is so highly polarised that the effect of rotating a piece of polaroid in front of the eyepiece of a refractor is marked. This does not work with a Newtonian. See Sky & Telescope 49, 21.

Nebula	RA	Dec	m^s	AD	T
Ced 194	21 37.7	+68 05	8.2	8	R
NGC 7133	21 44.0	+66 03	11.0	3	R
	22 34.8	+68 58		2	R
Sh 142	22 45.9	+57 57	Cluster	20 x 25	E

The cluster is NGC 7380.

Nebula	RA	Dec	m^s	AD	T
Sh 157	23 14.9	+59 51	9.1	3	E
	23 42.5	-15 26	var (R Aqr)	2 x 1	E

The nebula is excited by a small, hot companion to the Mira variable.

APPENDIX 1.

The Appendix contains supplementary observations of a variety of planetary nebulae by four observers. The first five pages are concerned with descriptions and drawings made by David Allen using 100, 60 and 40-inch reflectors. In two cases, M3-5 and M2-9, both descriptions and drawings are featured. Ten of the drawings are of objects which appear in the catalogue of planetary nebulae in Part One of this volume, while a number of the remainder are southern hemisphere objects. Orientation for these drawings is north up, east to the left.

Page 137 shows drawings of six nebulae; three of these are 18-inch observations, and three 6-inch. These six nebulae are also included in the catalogue of planetary nebulae, and are presented here to show the results achieved by different telescopes on the respective nebulae. Orientation for these objects is north down, east to the right.

Finally, on page 138, we show a low power rendering of the field of the planetary NGC 6803 as seen with an 8½-inch reflector used in conjunction with a 30° prism. Orientation here is north down, east to the right.

Appendix 1.

Cat	RA	Dec	m^{n}	AD	Con
M1-8	06 52.2	+03 14	11.6SBp	22	Mon
	Type: ?		RV ?		
	(60) Faint, 15" diam.; no central star.				
M3-4	07 54.1	-23 20	10SBr	14 x 3	Pup
	Type: IV		RV +64		
	(60) Faint, 20" - 25" diam.; no central star.				
M3-5	08 01.6	-27 36	8SBr	7 x 6	Pup
	Type: IV		RV +64		
	(40) Even illumination; star on NE edge.				
M2-9	17 04.4	-10 06	10SBr	29 x 10	Oph
	Type: VI		RV +91		
	(100) Blue star with fan of nebulosity that extends N - S about 15" each way.				
Na-1	17 11.5	-03 14	13.4	?	Oph
	Type: ?		RV +22		
	(100) Circular, 10" diam.; uniform green-blue with sharp edges; star just visible.				
M2-13	17 27.1	-13 24	?	10	Ser
	Type: ?		RV +77		
	(100) Blue star, suspected non-stellar.				
Hb 4	17 40.3	-21 41	13.6	6 x 5	Oph
	Type: IIIb		RV -61		
	(100) About 8" diam., slightly elliptical, no star; fluffy; PA 50° - 230°.				

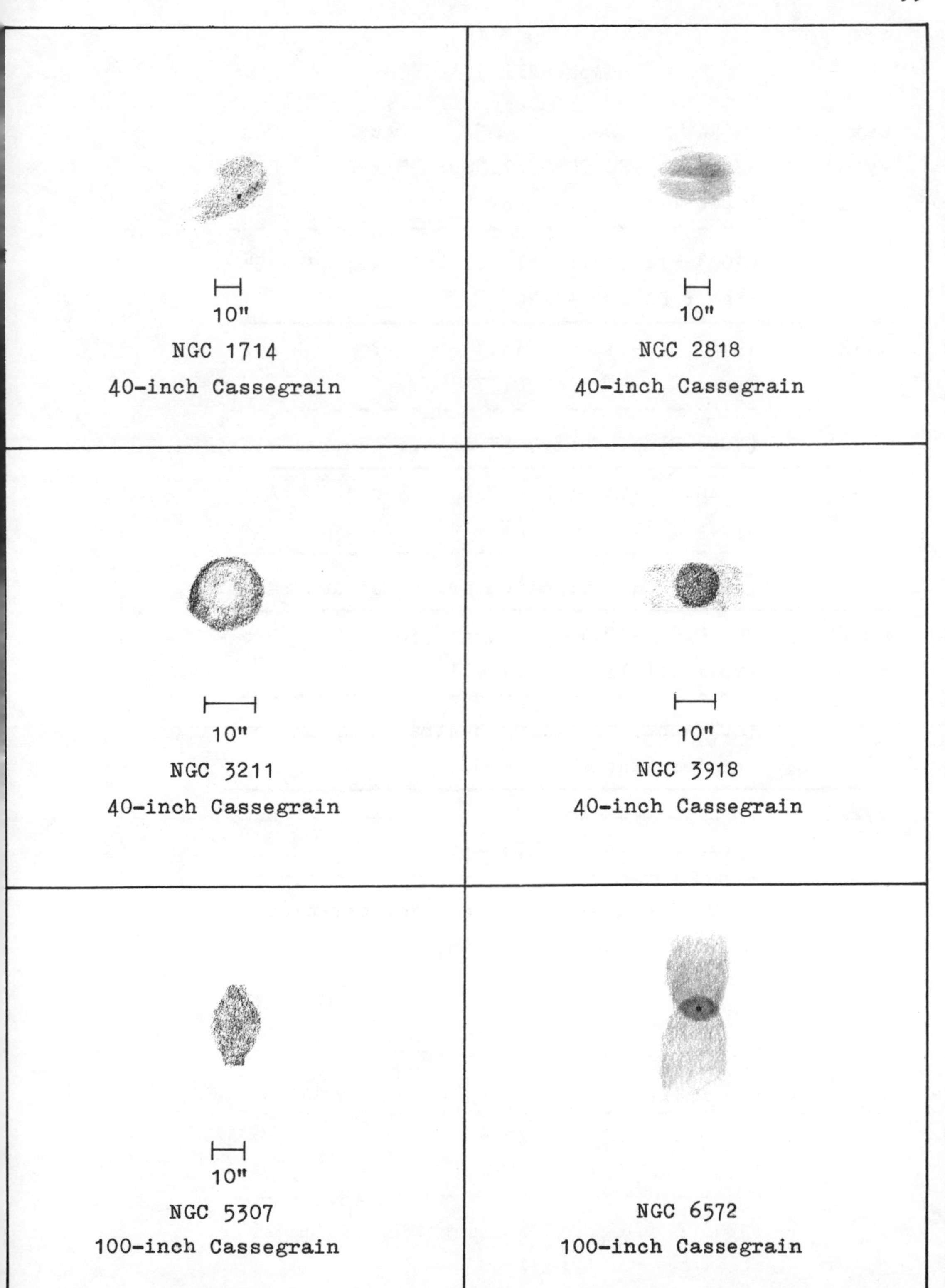
10"
NGC 1714
40-inch Cassegrain
10"
NGC 2818
40-inch Cassegrain
10"
NGC 3211
40-inch Cassegrain
10"
NGC 3918
40-inch Cassegrain
10"
NGC 5307
100-inch Cassegrain
NGC 6572
100-inch Cassegrain

Appendix 1.

Cat	RA	Dec	m^n	AD	Con
Vy1-2	17 53.3	+28 00	7.5SBr	5 x 4	Her
	Type: ?		RV ?		
	(100) Blue, slightly elliptical, 4" - 5" diam.; PA 40° - 220°.				
Na-2	19 17.3	-11 08	13.3	?	Aql
	Type: ?		RV ?		
	(100) Blue, fuzzy, 5" diam.; central star.				
M1-73	19 40.0	+14 54	?	5 x 4	Aql
	Type: ?		RV +11		
	(100) 3" diam.; circular, blue; central star.				
He1-5	20 10.8	+20 15	12SBr	30 x 29	Sge
FG Sge	Type: III+II		RV +21		
	(60) Annular; bright central star and a fainter star on E edge.				
Vy2-3	23 21.6	+46 46	13	4	And
	Type: ?		RV -55		
	(100) Green star at centre of circular nebula of about 10" diam.				

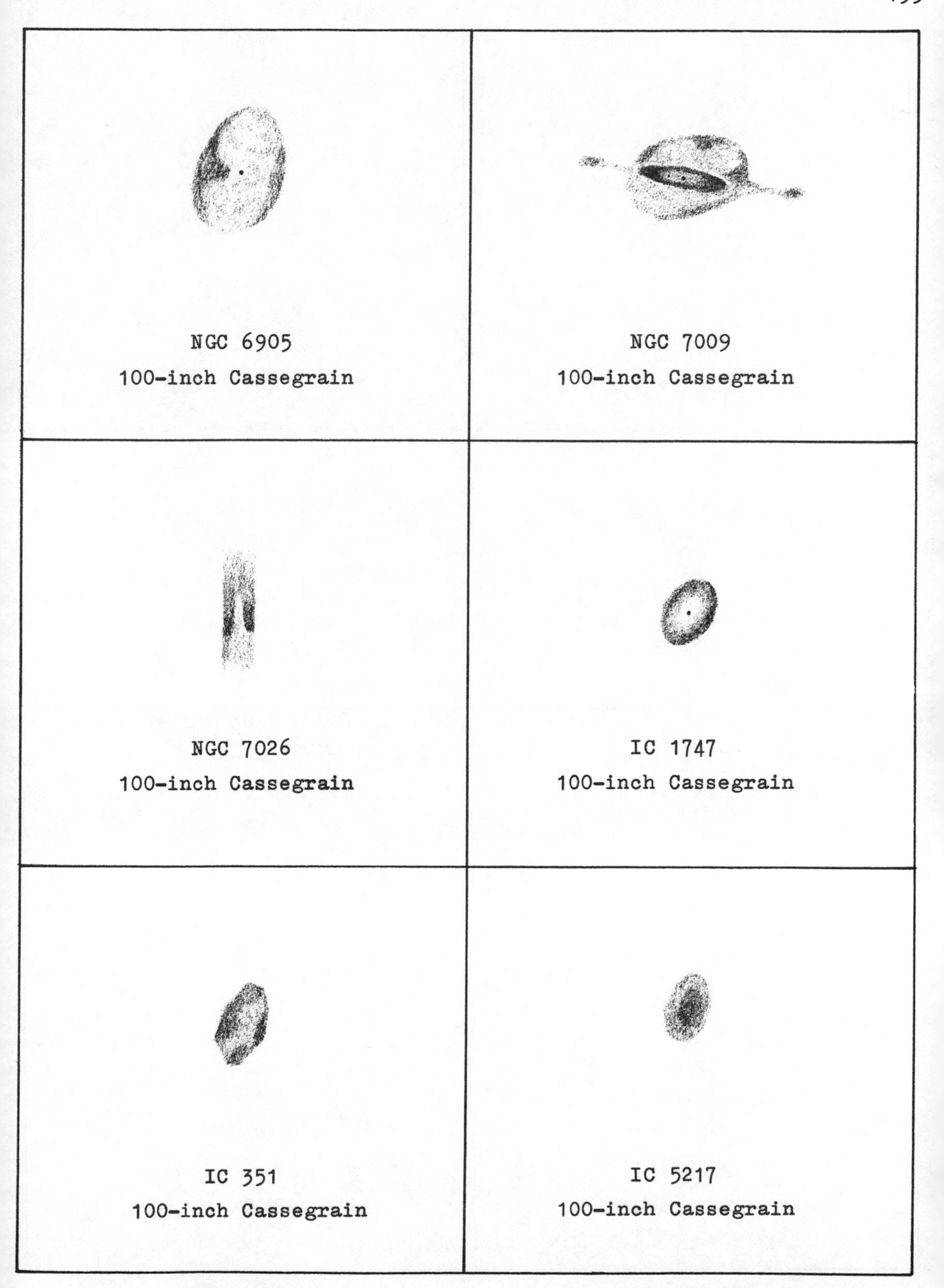

NGC 6905
100-inch Cassegrain

NGC 7009
100-inch Cassegrain

NGC 7026
100-inch Cassegrain

IC 1747
100-inch Cassegrain

IC 351
100-inch Cassegrain

IC 5217
100-inch Cassegrain

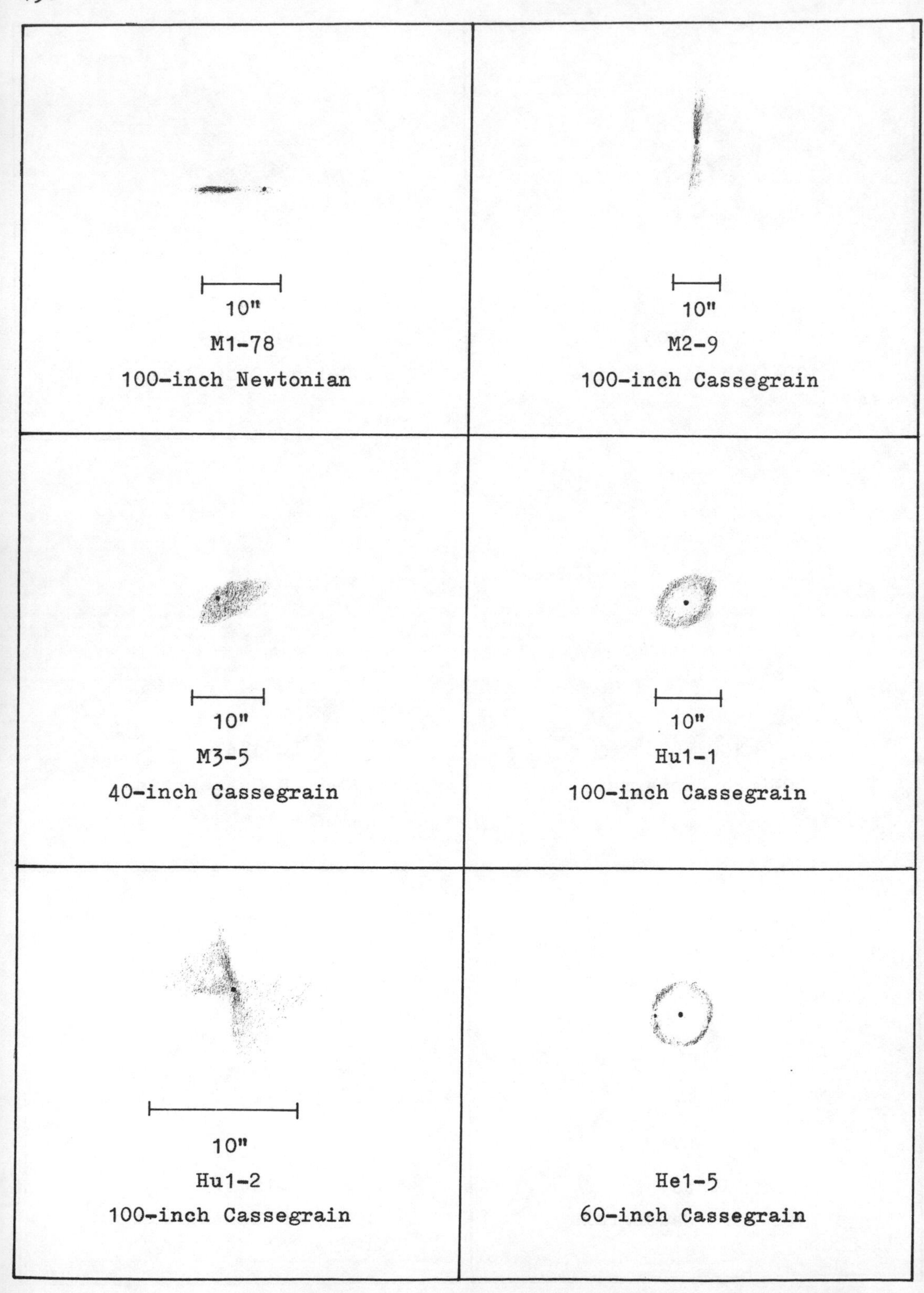
10"
M1-78
100-inch Newtonian
10"
M2-9
100-inch Cassegrain
10"
M3-5
40-inch Cassegrain
10"
Hu1-1
100-inch Cassegrain
10"
Hu1-2
100-inch Cassegrain
He1-5
60-inch Cassegrain

NGC 3242 18-inch J.K. Irving	NGC 6543 18-inch J.K. Irving
NGC 7293 18-inch J.K. Irving	NGC 650-1 M76 6-inch K. Sturdy
NGC 2440 6-inch K. Sturdy	NGC 7008 6-inch K. Sturdy

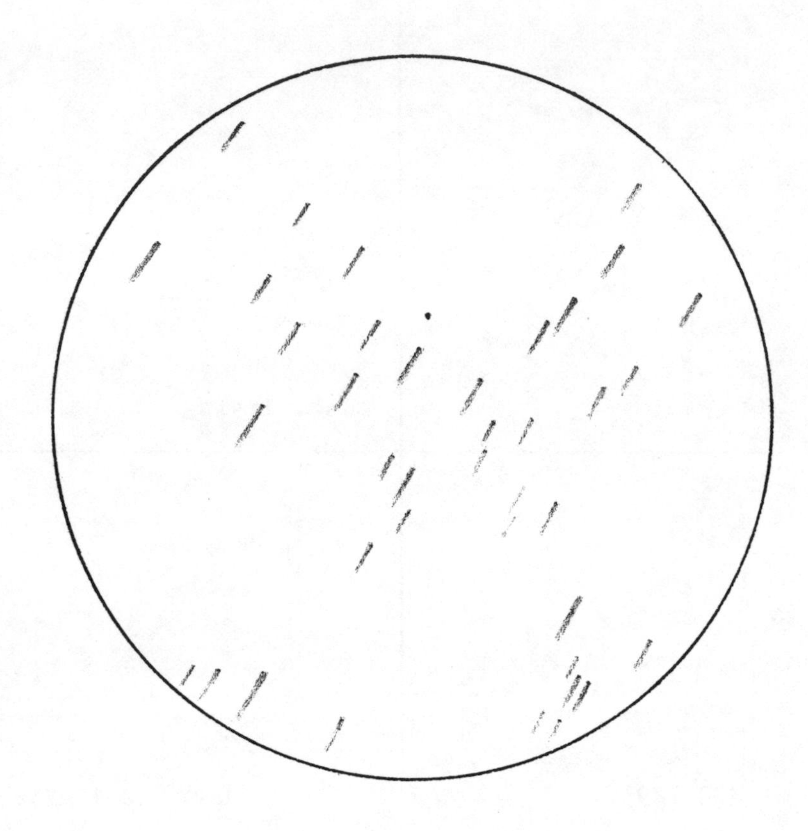

Field of the planetary nebula NGC 6803 as seen through a 30° prism applied to the eyepiece of an 8½-inch reflector. The emission spectrum of NGC 6803 is almost halfway between the centre and top edge of the field, which is 45' in diameter.

APPENDIX 2.

CLASSIFICATION OF PLANETARY NEBULAE: GURZADYAN'S SYSTEM.

The classification scheme for planetary nebulae devised by Gurzadyan is briefly described overleaf. Diagrammatic renderings of the respective classes are shown, plus examples of individual nebulae alongside each diagram.

Appendix 2.

The Gurzadyan classification system for planetary nebulae is, in many ways, a parallel scheme to that of Vorontsov-Velyaminov. A primary difference between the two systems can be seen in the diagram opposite of the Sp (spiral) class, exemplified by the planetary NGC 4361. In this object, the two slightly curved projections, which give it the Sp classification, are only to be seen on photographs in which the central regions of the nebula are heavily overexposed. Similarly, the classification of NGC 6826 as a double-envelope object is also at the expense of detail in the inner regions.

At the other end of the scale is the Sz (Z-like) classification of NGC 6778. On page 20 of this volume, the diagrammatic rendering of this planetary shows the Z-like structure as well as the mean boundaries of the outer regions. From this diagram it can be seen that the Z-like structure is present not in the outer, but in the inner regions of the nebula. Note also that two sections of the 'Z' form the brightest parts of the minor axis. This Z-like structure is present, in a greater or lesser degree, in a number of other planetaries, including M27.

From the above, it can be seen that a meaningful classification of certain planetaries cannot be from only inner or outer nebular distribution, but must be an amalgam of both. Thus the inner detail of NGC 4361 must be classified and added to the outer, while the reverse is true for NGC 6778. To take another example within the Gurzadyan classification, NGC 7662 is not only a ring nebula but also displays a double-envelope. The Gurzadyan classification is therefore an amalgam of these two structures, viz., III+II, and this method is, of course, also a feature of the classification system of Vorontsov-Velyaminov.

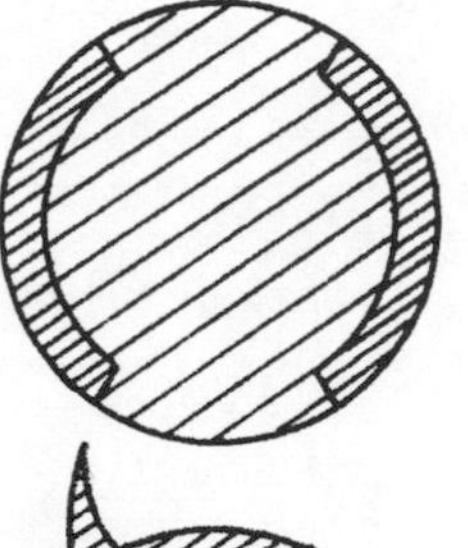

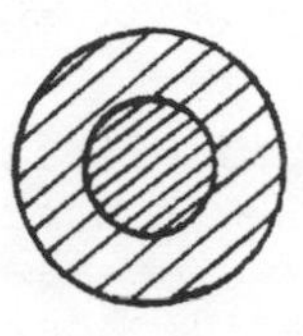

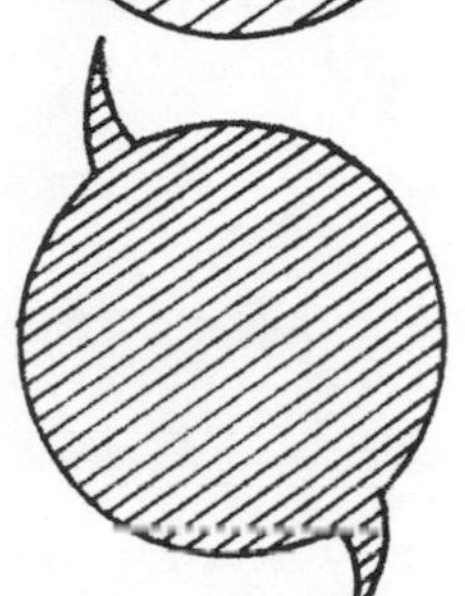

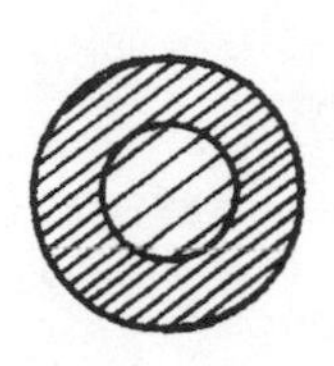

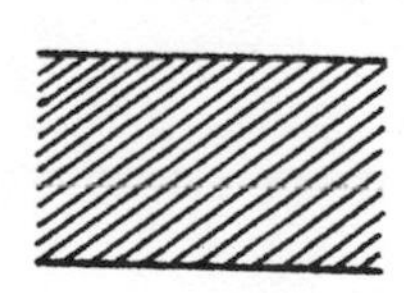

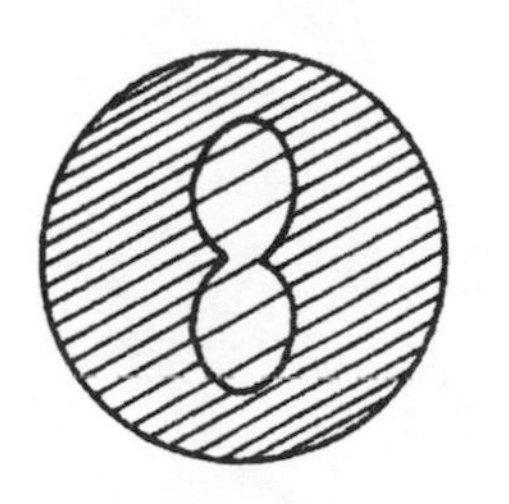

S Stellar IC 4997
D Diffuse NGC 2440

Alternative classification of Planetary Nebulae by Gurzadyan.

APPENDIX 3.

THE DISTANCES OF NEBULAE.

For historical reasons we have given estimates of the distances of the planetary nebulae but not of the gaseous nebulae. Nonetheless, the few available distances for the latter are generally much more reliable than those of the planetary nebulae.

Distance determination is one of the most difficult and yet most important aspects of astronomy. The only distances which are known to an accuracy that would be even remotely acceptable in a physics laboratory are those within the solar system. Parallax determinations are difficult because of the small amplitudes involved, and can be applied to only the very nearest stars. We may extrapolate to more distant star clusters if we assume that their fainter members have the same intrinsic luminosity as those stars of the same spectral types which are sufficiently nearby to have measurable parallaxes. Hence we derive intrinsic luminosities for hotter (and therefore brighter) stars, and use these to determine distances for more remote objects such as globular clusters and Cepheid variables. And so it goes on. But with each step more assumptions are made and more errors accumulate, so that when we try to give distances to extragalactic objects, we could easily be in error by a factor of two.

Although planetary and diffuse nebulae are within our galaxy, the determinations of their distances involve indirect methods and are therefore frought with additional sources of error. Even as bright and well-studies an object as M42 is a source of debate: based on observations of the fainter members of the star cluster associated with it, the published values of its distance range more than 20% either side of 500 pc.

For gaseous nebulae the following methods are available: (I) Spectroscopic. If a star or cluster is associated with the nebula, we may use its spectral type to predict its absolute magnitude, and from its apparent magnitude deduce the distance. Difficulties:
(a) The stars will usually be obscured to some extent by dust in the nebula. We must assume that the nebular dust has the same dimming properties as the general interstellar dust (itself not well-determined); this assumption is probably erroneous.
(b) Because the stars are in an H II region or reflection nebula they are young, and may not yet have settled onto the main sequence. If not, their absolute magnitudes will be wrongly chosen.
(II) Kinematic. Models of the galaxy, themselves based on the distance determinations mentioned above, can predict the radial velocity of an object for given coordinates and distance. This may be reversed to derive a distance from the measured radial velocity, although this is only practicable in certain parts of the sky. Difficulties: (a) The nebulae may have their own peculiar motions superimposed on that of the mean galaxy. (b) In many

Appendix 3.

cases (e.g., Cygnus, where we are looking along a spiral arm) two distances can produce the same velocity, and it may not be possible to choose one of these.

For planetary nebulae, the first method is impracticable. This is because their central stars are unusual objects whose absolute magnitudes are quite unpredictable. Moreover, planetary nebulae are not usually found in association with other, more conventional stars. The kinematic method is also unreliable because difficulty (a) becomes extreme. Reasonably accurate distances are known for only one planetary nebula, Pease 1, which is a member of the globular cluster M15. Distances of the few known planetary nebulae in external galaxies (mostly in the Large Magellanic Cloud) can also be inferred. All other determinations are little more than guesses.

APPENDIX 4.

HM Sge.

In the introduction to Part One of this Handbook, mention was made of the possible variability of the central stars of certain planetary nebulae. In particular, details were given of the variable star (FG Sge) which forms the nucleus of the planetary He1-5.

This latter object is located in a constellation which currently harbours another object of interest, the nebular variable, HM Sge.

In some ways this object surpasses FG Sge; certainly its rise in magnitude is spectacular, particularly when the short time scale involved is realised. During 1975, HM Sge showed a rise through 5 magnitudes in as many months (below 16 mag to 11 mag). Presently this object hovers at around the latter figure.

Regarded as a likely proto-planetary nebula, HM Sge shows spectral features characteristic of these objects. If this is a planetary in the making, then it presents an object of interest for visual observation. For those with direct-vision spectroscopes or prisms, this object can easily be picked out of the rich field it inhabits.

An 8½-inch reflector, used in conjunction with a prism, shows HM Sge well enough at low power. It must be pointed out that the visible spectrum of HM Sge is totally unlike that of those small planetaries which are usually found by means of spectroscope or prism. True, emission is seen, but not in the usual 4959 - 5007 Å [OIII] guise; instead, emission is seen within a bright stellar continuum. The identification is easily made, the emission line(s) appearing as a block of brighter luminosity near the centre of the spectrum of the star. Due to poor seeing during a number of observations, we were unable to confirm other, fainter, emission redward of the main emission. Possibly the extra dispersion of a direct-vision spectroscope would be able to clarify this, particularly when used in conjunction with a larger aperture.

The 1950 and 1975 positions of HM Sge are as follows:

1950	1975
$19^h\ 39^m.7\ +16^\circ\ 38'$	$19^h\ 40^m.8\ +16^\circ\ 41'$

APPENDIX 5.

PHOTOGRAPHIC SOURCES FOR PLANETARY AND GASEOUS NEBULAE.

In the majority of books or journals that the amateur astronomer is aquainted with, photographs of planetary and gaseous nebulae, if they are featured, are generally of the more well-known objects - M27, M57, M1, the Cygnus Loop, etc.

This leaves the amateur unfamiliar with interesting photographs of objects that appear not only in books, but also in professional astronomical journals. Accordingly we show below references relating to photographs of a selection of nebulae. The list does not pretend to be comprehensive, but should give a good indication of the different morphology of many of these nebulae.

Planetary Nebulae.

Curtis, H.D., Publ. Lick Obs., XIII, (1918).
Objects include NGC's 246, 1514, 6210, J 320.

Jeans, J., The Universe Around Us., Cambridge U.P., (1929).
NGC's 1501, 2022, 6720 (M57), 7662.

Evans, D.,
Thackeray, A., Mon. Not. Roy. Astr. Soc., 110, 429-439, (1950).
Objects include NGC's 3132, 3195, IC's 2448, 4723.

Aller, L.,
Liller, W., Stars and Stellar Systems, Vol. VII, Chap. 9. Univ. of Chicago Press, (1968).
Objects include NGC's 40, 2392, 6210, 6537, 7009, 7027, 7662, CD -29° 13998, M2-9.

Perek, L.,
Kohoutek, L., Catalogue of Galactic Planetary Nebulae., Czech. Academy of Sciences., Prague., (1968).
Many finding charts of very small nebulae which therefore show no detail. Larger-scale charts show detail in many objects, amon them: NGC's 2474-5, 7008, 7048, V-V 1-2, K1-16, He1-4.

Gurzadyan, G., Planetary Nebulae., Reidel, Dordrecht, (1970).
Objects include NGC's 4361, 6543, 6826, 7026, 7293, A65, A66, A70.

Louise, R., Astron. & Astrophys., 34, 21-22, (1974).
NGC 7662.

Peterson, A., Astron. & Astrophys., 53, 441-442, (1976).
Ps 1, planetary nebula in M15.

Appendix 5.

Gaseous Nebulae.

Thackeray, A., Mon. Not. Roy. Astr. Soc., 110, 524-530, (1950). Stars involved in nebulosity; also photograph of dark globules.

Dufay, J., Gaseous Nebulae and Interstellar Matter, Dover, (1968). Various nebulae.

Chopinet, M., et al., Astron. & Astrophys., 30, 233-240, (1974). Two nebulae, Sh2 - 255, Sh2 - 257.

Johnson, H.M., Astron. & Astrophys., 32, 17-19, (1974). NGC 7635.

Danziger, I.J., et al., Astron. & Astrophys., 37, 419-423, (1974). Large new southern reflection nebulae.

Glushkov, Yu I., Astron. & Astrophys., 39, 481-485, (1975). Star clusters in diffuse nebulosity.

Kohoutek, L., Wehmeyer, R., Astron. & Astrophys., 41, 451-453, (1975). V-V 1-7, probable reflection nebula.

Wendker, H.J., Smith, L.F., Astron. & Astrophys., 42, 173-185, (1975). NGC 6888.

Maucherat, A.J., Astron. & Astrophys., 45, 193-196, (1975). Sh2 - 106.

Deharveng, L., et al., Astron. & Astrophys., 48, 63-73, (1976). NGC 1491.

Danks, A.C., Manfroid, J., Astron. & Astrophys., 56, 443-446, (1977). NGC 6164-5.

Felli, M., et al., Astron. & Astrophys., 59, 43-52, (1977). NGC 2175.

Davies, R.D., Meaburn, J., Astron. & Astrophys., 69, 443-444, (1978). New optical supernova remnants.

APPENDIX 6.

BIBLIOGRAPHY.

Planetary Nebulae.

Abell, G.	PASP 67, 258.		1955
Abetti, G. Hack, M.	Nebulae and Galaxies. (Tr. by V. Barocas).	Faber & Faber	1963
Aller, L.H.	PASP, 575, 574.		1976
Becvar, A.	Atlas Coeli Katalog.	Czech. Acad. of Sciences.	1964
Berman, L.	Astr. J. 60, 151.		1955
Cahn J. Kaler, J.B.	Ap. J. Supp. 22, 319.		1971
Curtis, H.D.	Pub. Lick Obs. XIII.		1917
Glyn Jones, K.	Messier's Nebulae and Star Clusters.	Faber & Faber.	1968
Gurzadyan, G.A.	Planetary Nebulae.	Gordon & Breach.	1969
Longmore, A.J.	Mon. Not. Roy. Ast. Soc. 178, 251.		1977
O'Dell, C Osterbrock, D. (Eds.).	Planetary Nebulae. (I.A.U. Symp. 34)	Reidel.	1968
Perek, L. Kohoutek, L.	Catalogue of Galactic Planetary Nebulae.	Czech. Acad. of Sciences.	1969
Vorontsov-Velyaminov, B.A.	New Catalogue of Planetary Nebulae.	Moscow.	1962
Wilson, O.C.	Ap. J. 108, 201.		1948

Gaseous Nebulae.

Becvar, A.	Atlas Coeli and Catalogue.	Sky Pub.Corp.	1973
Lynds, B.T.	Ap. J. Supp. 7, 1.		1962
	Ap. J. Supp. 12, 163.		1965
Rodgers, et al.	Mon. Not. Roy, Ast, Soc. 121, 103.		1960
Sharpless, S.	Ap. J. Supp. 4, 257.		1959